House Wiring

by Roland E. Palmquist

THEODORE AUDEL & CO.
a division of
THE BOBBS-MERRILL CO., INC.
Indianapolis/New York

FIFTH EDITION

FIRST PRINTING—1982

Library of Congress Cataloging in Publication Data

Palmquist, Roland E.
  House wiring.

  1. Electric wiring, Interior. I. Title.
TK3285.P34   1982   621.319'24   81-21674
ISBN 0-672-23364-9   AACR2

Foreword

With the establishment of a *Residential Journeyman Wireman Classification*, it was felt that a book covering this subject might be very helpful. As an inspector, I find that many journeymen who work mostly with industrial and commercial types of wiring may occasionally be required or desire to wire a residence. It was felt that, for these reasons, a book covering only residential wiring would be quite appropriate.

The importance of a good safe residential wiring job is sometimes taken too lightly. Poorly installed wiring can be very hazardous to the owner or renter as well as to his personal property. A good and safe wiring installation will not only be a source of satisfaction to the electrician, but to the owner or renter as well.

Electricity is too often taken as a matter-of-fact item when, in fact, it is one of our most important servants. We use it without giving it a second thought, and we are inclined to completely ignore the wiring installation as long as the lights come on when we push a switch.

My work in the electrical industry dates back to 1928. I often look in wonderment at the changes that have taken place in this field since that time. Then, in most cases, one or two branch circuits served the entire residence. Today, nearly everything around and in the home depends on electricity. This includes housekeeping chores, entertainment, hobbies, heating, cooling—the list is endless.

The average person spends the major portion of his time in his home. Most of his worldly goods are there, his loved ones are there, so every precaution should be taken to insure that safe and adequate electrical wiring is installed. Planning for the future growth of electrical usage should also be done. This will involve only a small amount of extra money if done at this time, but if the customer should later find the wiring inadequate for his needs, then many dollars will be required to bring it up to date. This fact can be seen from the tremendous increase in electrical consumption since 1928 when one or two branch circuits were adequate. In the home of today, twenty or more circuits are not uncommon. Electrical usage has far exceeded the predictions of the electric utility companies.

With all of this in mind, I would like to dedicate this book to the men who have given their time and money to develop the NATIONAL ELECTRICAL CODE to where it is today. Among the many persons involved, I especially want to mention Mr. George H. Tryon, former NFPA Director of Membership Services; Mr. Richard Lloyd, former Chairman of the Correlating Committee; Mr. Frank Stetka, former Secretary and Field Engineer; Mr. John Watt, former Secretary and Field Engineer; and Mr. L. E. LaFehr, former Member of the Correlating Committee and Director of the International Association of Electrical Inspectors. These men have graciously permitted me to use excerpts from the NATIONAL ELECTRICAL CODE in the preparation of this book.

Every attempt has been made to offer a clear interpretation of the NEC 1981 rulings and to anticipate changes that are likely to be incorporated in the 1981 NEC. Consult the 1981 NEC when it is available for any changes that have not been included.

ROLAND E. PALMQUIST

Contents

CHAPTER 1

Introduction

We are all familiar with the NATIONAL ELECTRICAL CODE published by the National Fire Protection Association, 470 Atlantic Avenue, Boston, Mass. 02210. This Code has been adopted by inspection authorities throughout the country. There are two editions — the full length edition known as the *NFPA 70*, and the ELECTRICAL CODE FOR ONE AND TWO FAMILY DWELLINGS known as NFPA 70A. Both are available from the above address, or may be secured from your local inspection authority. It is highly recommended that one or both be used in conjunction with this book.

Code language is a special language, so become thoroughly familiar with it. Some people complain it is difficult to find something in the Code, so use this book along with the Code and familiarize yourself with it. This book was prepared in a step-by-step manner with many illustrations added to help simplify and clarify the Code rulings. It is impossible to include every conceivable answer for every conceivable condition, but it should be possible to piece together special conditions. The understanding gained here may be supplemented by asking the inspector about points on which you are in doubt. In most every case he will be able to assist you with any specific problems that might arise. Be assured that an inspector's job is not to lay out the job for you. He is too busy, but he will most certainly assist you in your specific problems as they arise.

Most states, many counties, and practically every city or town have laws governing the installation of electrical wiring. They also require permits, licenses, and inspections. The first step is to ascertain under which jurisdiction the inspections will come. Next, obtain a permit (if one is necessary) and pay the fees. Call for all inspections, such as the rough-in inspection before closing in the walls, and the final inspection. The inspector might well require walls to be removed should they be covered before inspection. He will actually be doing you a favor in doing so since he cannot give an honest inspection if he can't see the wiring as installed before covering. You will find that the inspector has only one job to do—to see that there are no hazards to you and to the customer's home and family. Should a fire result without proper inspection, there could be a possible involvement with the collection of insurance claims and responsibility. Anyone purchasing or renting a home has a right to demand proof that the wiring has been inspected. They expect to find an approval tag on it.

Actually, the wiring of a home is the smallest single cost of the total construction costs. This being the case, do not slight the smallest detail of the wiring installation.

As you progress with the reading of this book, you will find references to certain portions of the Code, such as **Section 230-70** or **Table 370-6(a)**. These indicate the sections or Tables as numbered in the NATIONAL ELECTRICAL CODE (NEC) so that, should you wish to do so, you may look them up in the NEC for further study. There will also be quotes from the NEC as well as some Tables. The quotes will be in italics and the Tables identified by Notes.

Section 90-1 in **Article 90** of the NEC explains the purpose of the NEC:

90-1. Purpose.

(a) **Safeguards.** *The purpose of this Code is the practical safeguarding of persons and property from hazards arising from the use of electricity.*

(b) **Adequacy.** *This Code contains provisions considered necessary for safety. Compliance therewith and proper maintenance will result in an installation essentially free from hazard, but not necessarily efficient, convenient, or adequate for good service or future expansion of electrical use.*

(c) **Intention.** *This Code is not intended as a design specification nor as an instruction manual for untrained persons.*

> *Hazards often occur because of overloading of wiring systems by methods or usage not in conformity with this Code. This occurs because initial wiring did not provide for increases in the use of electricity. An initial adequate installation and reasonable provisions for system changes will provide for future increases in the use of electricity.*

Bearing in mind what is covered in **Section 90-1**, you will find many recommendations added to this book as good practices. They will be over and above Code requirements and will be so stated along with reasons for the recommendations. Consider putting these recommendations into use. You will be surprised at how easy the extras are to sell to the customer, and they will pay dividends for a long time to come.

We must, by necessity, have definitions for clarification. **Article 100** of the NEC is all definitions. Definitions will not

9

appear in this book as a separate chapter, but will be presented as it is felt they are applicable. In this way you will receive the most benefit from these definitions because you will see the application as the particular subject is being covered.

Some of the interpretations will be taken from *Audel's* GUIDE TO THE 1981 NATIONAL ELECTRICAL CODE, written by this author. Remember that this book is based on the NEC and that some states, counties, cities, or towns may have additional requirements that must be followed in the wiring installations in those jurisdictions. For instance, Colorado has adopted the NEC. This permits wiring residences with nonmetallic cable (Romex). There are, however, some cities in Colorado which require metallic raceways for these wiring installations and do not permit NM cable. There are other requirements in other cities, so check with the authority having jurisdiction for any requirements over and above the NEC requirements. The NEC, as was stated in **Section 90-1**, is a minimum. However, be assured that with all of the dedicated and highly qualified men participating in the writing of the NEC, that whenever they find something has been included that tends toward a hazard, the Code is revised. In the meantime, if the methods are safe, they do not wish to penalize persons by requiring wiring methods not deemed necessary. They, of course, cannot designate fire-zones, etc., so these must be left up to the local inspection authorities.

CHAPTER 2

Basis for Load Calculations

In order to plan the wiring, it will be necessary to do a certain amount of calculating for the Service, Branch Circuits, and Feeder Circuits. The definitions for these three terms as they appear in *Audel's* GUIDE TO THE 1981 NATIONAL ELECTRICAL CODE are as follows.

Service—*The conductors and equipment for delivering energy from the electricity supply system to the wiring system of the premises served.* This definition is very complete and applies to all wiring and equipment extending from the last pole or underground distribution system through the service equipment. The following definitions will give the breakdown of the separate parts or sections of a service.

Service Conductors—*The supply conductors which extend from the street main, or from transformers to the service equipment of the premises supplied.* Therefore, service conductors are the conductors defined under **Service.**

Service Cable—The service conductors in the *form of a cable.*

Service Drop—*The overhead service conductors from the last pole or other aerial support to and including the splices, if any, connecting to the service-entrance conductors at the building or other structure.* In rural areas, the utility company often locates a meter pole in the yard; this may or may not have an overcurrent device installed. The service drop does not stop at the pole, but continues on to the building or buildings or other structures that it serves. See Fig. 1.

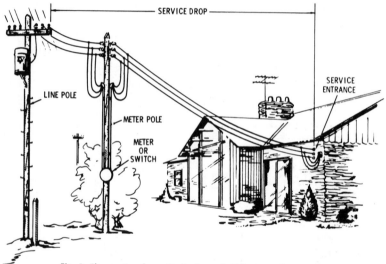

Fig. 1. The service drop attached to a building or other structure.

Service Entrance Conductors, Overhead System— The service entrance includes the conductors from the service equipment to a point outside the building, clear of the building walls. They are attached to the service drop at this point by either tap or splice. The meter housing and meter, if on the building wall, are not considered as parts of the service-entrance equipment.

Service Entrance Conductors, Underground System
—*The service conductors between the terminals of the service equipment and the point of connection to the service lateral. Where service equipment is located outside the building walls, there may be no service-entrance conductors, or they may be entirely outside the building.* See Fig. 2 for a sketch showing the possible conditions.

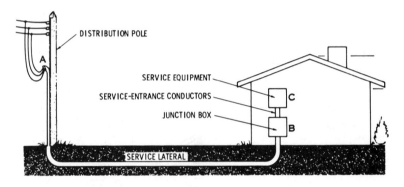

Fig. 2. The service lateral extends from point A to point B. The service entrance is from point B to point C.

Service Equipment—The necessary equipment usually consists of a circuit breaker or fuses and a switch located on the inside or outside of the building near the point of entrance. It is intended to constitute the means of disconnecting the electrical supply entering the building, as shown in Fig. 3. Further information is covered in **Article 230.**

Service Lateral—These include the underground service conductors, including any risers up the pole at the street main or transformer structure. They are considered as service laterals until they enter a junction box

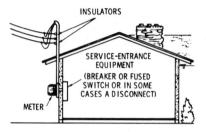

INSULATORS

SERVICE-ENTRANCE
EQUIPMENT
(BREAKER OR FUSED
SWITCH OR IN SOME
CASES A DISCONNECT)

METER

Fig. 3. Service-entrance equipment that serves as the electrical disconnect supply.

in the building. If such a box is not used, they will cease to be service laterals at the point of entrance into the building, at which they become service-entrance conductors. If the service-entrance equipment is located on the outside of the building, there may possibly be no service-entrance conductors; they could all be termed **service laterals.** See Fig. 3.

Service Raceway—This is any raceway, conduit, or tubing enclosing the service-entrance conductors. Where a service mast is used, the conduit to the metering cir-

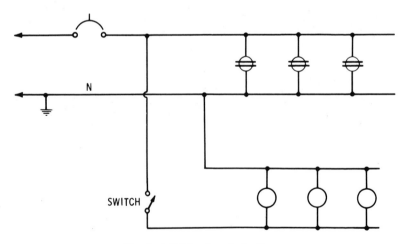

N

SWITCH

Fig. 4. A lighting branch circuit.

cuit, the raceway (for connecting to the metering if several should be required), and the connections from the raceway to the service equipment are all considered the service raceway.

Branch Circuits—There are five separate definitions for branch circuits, each with an individual purpose in mind.

(1) *A branch circuit is that portion of a wiring system extending beyond the final overcurrent device protecting the circuit.* This protective device may be a circuit breaker or fuse in the service-entrance equipment, where the particular circuit breaker or fuse does not serve as protection for feeder circuits to the feeder panels, or where feeder panels are used. The branch circuit is that circuit coming from the feeder panel, or as stated above, it may be from the service-entrance equipment.

Do not confuse the part that says *beyond the final overcurrent device protecting the circuit.* The part *protecting the circuit* defines what is meant. Thus, in the lighting-branch circuits shown in Fig. 4, there is a circuit breaker or fuse in the panel feeding the lighting-branch circuit.

(2) A branch circuit (as applied to appliances) is a circuit designed for the purpose of supplying an appliance or appliances; nothing else can be connected to this circuit, including lighting. The lighting that is an integral part of the appliance is not considered as lighting in this instance. Pay particular attention to this part, as we will be covering appliance circuits as they apply to residential wiring.

(3) A branch circuit multiwire (Figs. 5A and 5B) is a circuit consisting of two or more ungrounded con-

ductors with an equal potential between them, and a grounded conductor with an equal potential between it and any one ungrounded conductor. In residential wiring, this is a 120/240-volt system in most cases, but it

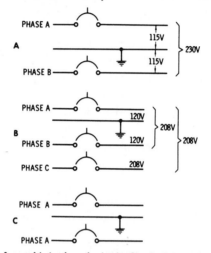

Fig. 5. Variations of a multiwire branch circuit. Circuit C is not a multiwire branch circuit because it utilizes two wires from the same phase in conjunction with the neutral conductor.

may also be a 120/208-volt wye system. Fig. 5C is not a multiwire branch circuit because it utilizes two ungrounded conductors from the same phase in conjunction with the neutral conductor.

(5) A branch circuit individual is a circuit that supplies just one piece of equipment, such as a motor, an air conditioner, or a furnace.

Only four out of five definitions are given here as it is not felt that the other one particularly applies to residential wiring.

Feeder—*All circuit conductors between the service equipment, or the generator switchboard of an isolated* *plant, and the final branch-circuit overcurrent device.* We must always have service equipment, but it is not always necessary to have feeders. For instance, the average home does not have feeders; the branch circuits are taken from the overload devices in the service-entrance equipment. When we have a large area to cover, the usual practice is to extend feeder circuits from the service-entrance equipment to the proper locations for distribution to the branch circuits. Thus, the conductors from the service equipment to the distribution location are termed *feeders.*

In the preceding definitions, the NEC quotations appear in *italics.*

For the information necessary to arrive at the calculations of loads, we will use 3 watts per square foot for residential occupancies. There will be some cases where this figure will not be sufficient, such as when more lighting than normal is installed in a residence. For general calculations, however, 3 watts per square foot will be adequate. Should there be extra lighting, this will be added to the total obtained from the 3-watts-per-square-foot figure.

In figuring the watts per square foot for a residence, the outside dimensions of the building shall be used. These dimensions do not include the area of open porches or garages which are attached to the residence. If there is an unfinished basement, it should be assumed that it will be finished at a later date. Thus, the square foot area of the basement should be included in the square foot area used for load calculations so that the system will be adequate when the basement is finished.

In Figs. 6A and 6B, we find the first floor measures 32½ ft. by 57 ft., or 1853 square feet. The basement is only a partial basement, the measurements of which are 32½ ft. by

17

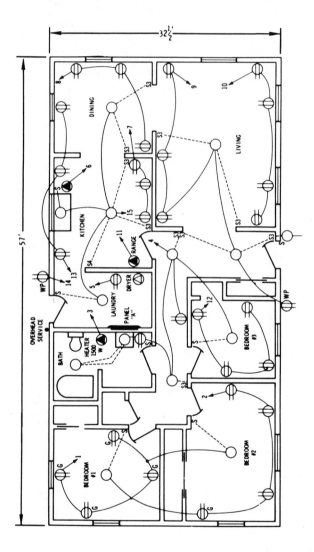

(A) First-floor layout.

Fig. 6. Electrical layout drawing of

18

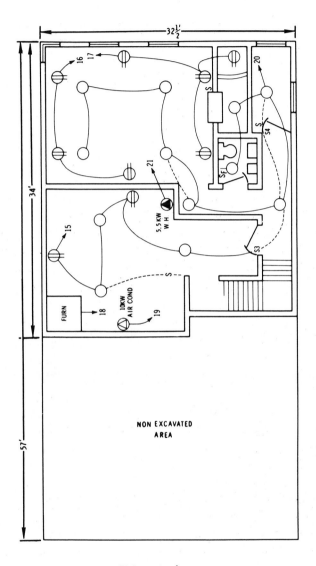

(B) Basement layout.

a one-story residence with basement.

34 ft., or 1105 square feet. From this we find a total of 2958 square feet which will be used with the 3-watts-per-square-foot figure for general lighting, etc. We find that this comes to 8874 watts total.

The above watts per square foot is a general lighting load. Any outlet of 15 amperes or less, such as receptacles in bedrooms, living rooms, bathrooms, etc., are considered in the general lighting load. Any loads other than those considered as general lighting loads, such as special lighting, heating, air cooling (a special loading is considered for heating and air cooling, and will be covered later), or any special motor loads, etc., will be considered separately. Ranges and electric dryers will also be considered separately.

In general, for illumination as covered under watts per square foot in dwelling occupancies, it is recommended that not less than one branch circuit be installed for each 500 square feet of floor area in addition to that required for special loads.

Section 220-3(b) covers what are termed small appliance branch circuits. These are not to include any appliances that are fixed, such as disposals and built-in dishwashers, etc. To quote **Section 220-3(b)**:

(b) Small Appliance Branch Circuits, Dwelling Unit . For the small appliance load in kitchen, pantry, family room, dining room, and breakfast room of dwelling occupancies, two or more 20-ampere appliance branch circuits in addition to the branch circuits specified in **Section 220-3(a)** shall be provided for all receptacle outlets in these rooms, and such circuits shall have no other outlets.

Receptacle outlets supplied by at least two appliance receptacle branch circuits shall be installed in the kitchen.

At least one 20-ampere branch circuit shall be provided for laundry receptacle(s) required in **Section 220-3(c)**.

ADDITIONS TO EXISTING INSTALLATIONS

Section 220-2(**b** or **d**) apply to new circuits or extensions to existing electrical systems in dwellings. They may be figured on the watts-per-square-foot basis or the amperes-per-outlet basis. This will apply to that portion of an existing building that has not been previously wired, or to any addition that exceeds 500 square feet in area. In this case, the addition will be figured on the watts-per-square-foot basis, as previously described. Also take into consideration any other loads that are involved.

SMALL APPLIANCE LOADS

The small appliance branch circuits just mentioned shall have a feeder load of not less than 1500 watts for each two-wire circuit installed, as outlined in **Section 220-3**(**b**). These circuits are for small appliances only (portable appliances supplied from receptacles of 15- or 20-ampere rating). Recall that a minimum of two such 20-ampere (1500-watt) circuits shall be installed in the kitchen. It will often be desired to install more than the required minimum of two. Remember that small appliance circuits are also required in the pantry, family room, dining room, and breakfast room. These may be fed from the two required in the kitchen or separate 1500-watt (20-ampere) circuits may be installed. These 20-ampere small appliance circuits and the 20-ampere laundry circuit all must be installed with No. 12 copper conductors or No. 10 aluminum conductors.

What constitutes a family room is not clearly defined. It is not a living room or parlor, but is a room where the family will gather to relax, play games, have snacks, look at television, etc. The reason for small appliance circuits in this room is because snack-type meals or popcorn may often be prepared or served there, so 20-ampere receptacle cir-

21

cuits are installed to handle the plug-in type portable electrical appliances which may be used in this room from time to time without overloading the normal circuits. The important point to remember is not what this room is called, but what it will be used for.

ELECTRIC RANGES

In calculating feeder loads for electric ranges or other cooking appliances in dwelling occupancies, any that are rated over 1¾ KW shall be calculated according to **Table 220-19** in the NEC. Some of the notes accompanying the table are a part thereof and are important in the calculation of feeder and branch circuits.

Due to larger wattages being used in modern electric ranges, it is recommended that the minimum demands for any range less than 8¾ KW rating be figured using Column A in **Table 220-19.**

The notes accompanying this table do not appear here. Should you need to refer to them, see NFPA 70 or 70A.

It is wise when using this table not to merely size the branch circuit and/or feeder circuit to the cooking unit(s) installed according to this table, but to have additional capacity in the conductors so that if a higher wattage range is added later, it will not be necessary to rewire the circuit. The additional cost at the time of the original installation will be very small.

Demand factor, as in the table, is defined as follows:

Demand Factor—Demand factor may be applied to an entire electrical system, or any part of an electrical system. It is the *ratio of the maximum demand of the system, or part of a system, to the total connected load of the system, or of the part of the system under con-*

Table 220-19. Demand Loads for Household Electric Ranges, Wall-Mounted Ovens, Counter-Mounted Cooking Units and Other Household Cooking Appliances over 1 ¾ kW Rating

NUMBER OF APPLIANCES	Maximum Demand (See Notes) COLUMN A (Not over 12 kw Rating)	Demand Factors (See Note 3)	
		COLUMN B (Less than 3½ kw Rating)	COLUMN C (3½ kw to 8¾ kw Rating)
1	8 kw	80%	80%
2	11 kw	75%	65%
3	14 kw	70%	55%
4	17 kw	66%	50%
5	20 kw	62%	45%
6	21 kw	59%	43%
7	22 kw	56%	40%
8	23 kw	53%	36%
9	24 kw	51%	35%
10	25 kw	49%	34%
11	26 kw	47%	32%
12	27 kw	45%	32%
13	28 kw	43%	32%
14	29 kw	41%	32%
15	30 kw	40%	32%
16	31 kw	39%	28%
17	32 kw	38%	28%
18	33 kw	37%	28%
19	34 kw	36%	28%
20	35 kw	35%	28%
21	36 kw	34%	26%
22	37 kw	33%	26%
23	38 kw	32%	26%
24	39 kw	31%	26%
25	40 kw	30%	26%
26-30	{15 kw plus 1 kw for each range}	30%	24%
31-40		30%	22%
41-50	{25 kw plus ¾ kw for each range}	30%	20%
51-60		30%	18%
61 & over		30%	16%

sideration. The loads on a system are practically never thrown on at the same time due to the diversity of uses. Somewhere between the maximum connected load and the actual usage is a load that may be considered the maximum demand. This fact is often used in determining the size of conductors or overcurrent devices. The demand factor is usually determined by a series of tests, and then after it is proved it is added to the Code. Examples will appear later.

As a rule, the entire wattage of the range will not be used at one time, and this is the reason that demand factors have been taken into consideration in **Table 220-19.**

CLOTHES DRYERS (ELECTRIC)

Electric clothes dryers will be taken at 100% of the name-plate rating. As with ranges, it is a good policy to install the branch circuit and feeder conductors a little larger than the minimum required. I recommend nothing less than No. 8 copper or No. 6 aluminum to the dryer. Some wiremen have a tendency to install conductors that barely meet the name-plate rating, but with higher wattage dryers appearing, it certainly is false economy not to install sufficiently large conductors in the original installation.

SPACE HEATING AND COOLING

In dwelling occupancies having both electrical space-heating and air-cooling equipment, the larger of the two loads is used in calculations and the smaller omitted, providing that the likelihood of both being used at the same time is remote. An air conditioning load is an inductive-type load, so the number of amperes drawn by this equipment is used in the calculations in order to take into account the low power factor involved.

FARM BUILDINGS

Many residences will be on farms. There are calculations for farm buildings which will appear in a separate chapter.

CALCULATIONS OF FEEDER LOADS

The following excerpt from **Table 220-11** applies to dwellings:

Table 220-11. Lighting Load Feeder Demand Factors

Type of Occupancy	Portion of Lighting Load to which Demand Factor Applies (wattage)	Feeder Demand Factor
Dwelling Units	First 3000 or less at Next 3001 to 120,000 at Remainder over 120,000 at	100% 35% 25%
*Hospitals	First 50,000 or less at Remainder over 50,000 at	40% 20%
*Hotels and Motels — including Apartment Houses without provision for cooking by tenants	First 20,000 or less at Next 20,001 to 100,000 at Remainder over 100,000 at	50% 40% 30%
Warehouses (Storage)	First 12,500 or less at Remainder over 12,500 at	100% 50%
All Others	Total Wattage	100%

* The demand factors of this Table shall not apply to the computed load of sub-feeders to areas in hospitals, hotels and motels where entire lighting is likely to be used at one time; as in operating rooms, ballrooms, or dining rooms.

OPTIONAL CALCULATION

The following is quoted from the NEC as it applies to dwellings:

220-30. Optional Calculations—Dwelling Unit.

(a) **Feeder and Service Load.** *For a dwelling unit having the total connected load served by a single 3-wire, 115/230-volt or 120/208-volt set of service-entrance conductors or feeder conductors with an ampacity of 100 or greater, it shall be permissible to compute the feeder and service loads in accordance with Table 220-30 instead of the method specified in Part B of this Article. Feeder and service-*

entrance conductors whose demand load is determined by this optional calculation shall be permitted to have the neutral load determined by **Section 220-22.**

(b) **Loads.** *The loads identified in* **Table 220-30** *as "other loads" and as "Remainder of other load" shall include the following:*

(1) *1500 watts for each 2-wire, 20-ampere small appliance branch circuit and each laundry branch circuit specified in* **Section 220-16.**

(2) *3 watts per square foot for general lighting and general-use receptacles.*

(3) *The nameplate rating of all fixed appliances, ranges, wall-mounted ovens, counter-mounted cooking units, and including four or more separately controlled space heating units.*

(4) *The nameplate ampere or kVA rating of all motors and of all low-power-factor loads.*

(5) *When applying* **Section 220-21** *use the largest of the following: (1) Air-conditioning load; (2) The 65% diversified demand of the central electric space heating load; (3) The 65% diversified demand of the load of less than four separately controlled electric space heating units; (4) The connected load of four or more separately controlled electric space heating units.*

Table 220-30. Optional Calculation for Dwelling Unit

Load in kW or kVA	Demand Factor Percent
Air conditioning and cooling, including heat pump compressors	100
Central electric space heating	65
Less than four separately controlled electric space heating units	65
First 10 kW of all other load	100
Remainder of Other load	40

220-31. Optional Calculation For Additional Loads in Existing Dwelling Unit.

Quoting from the NEC:

For an existing dwelling unit presently being served by an existing 120/240 volt or 208Y/120, 3-wire, 60-ampere service, it shall be permissible to compute load calculations as follows:

Load (kW or kVA)	Percent of load
First 8 kW of load at	100%
Remainder of load at	40%

Load calculation shall include lighting at 3 watts per square foot; 1500 watts for each 20-ampere appliance circuit; range or wall-mounted oven and cooking unit, and other appliances that are permanently connected or fastened in place, at nameplate rating.

If air conditioning equipment or electric space heating equipment is to be installed the following formula shall be applied to determine if the existing service is of sufficient size.

Air conditioning* 100%

Central electric space heating* 100%

Less than four separately controlled space
*   heating units* 100%*

First 8kW of all other load 100%

Remainder of all other load 40%

Other loads shall include:

1500 watts for each 20-ampere appliance circuit

Lighting and portable appliances at 3 watts per sq. ft.

Household range or wall-mounted oven and counter-mounted cooking unit.

   **Use larger connected load of air conditioning and space heating, but not both.*

All other fixed appliances including four or more separately controlled space heating units, at nameplate rating.

MOTOR LOADS

The NEC will be referred to:

Section 220-14. Motors. *Motor loads shall be computed in accordance with* **Sections 430-24, 430-25, and 430-26.**

The basic discussion here is that if the motor loads are included in the residence (and this will include air conditioning motors, furnace motors, pump motors, disposal motors, etc.), 125% of the full-load current rating of the largest motor involved will be taken, and then 100% of the full-load current rating of the balance of the motors used in determining the ampere rating of the conductors required to serve the service and/or feeders in the residence.

CHAPTER 3

Actual Calculations for Dwellings

The first calculations in this Chapter will be on the basis of **Table 220-11** of the NEC and Chapter 2 of this book. This method is widely used. The heading of **Table 220-11** is: **Lighting Load Feeder Demand Factors.** Do not let this confuse you, it is for Feeder Calculations and/or Services.

CALCULATION NO. 1

Let us use the dimensions of the residence illustrated in Fig. 6A and 6B of Chapter 2. The outside dimensions of the ground floor are 32½ ft. by 57 ft., or 1853 sq. ft. The basement was (in round figures) 32½ ft. by 34 ft., or 1105 sq. ft. Dimensions less than 6 inches and small offsets can be ignored as they do not materially influence the final calculations. Assume that the basement will be finished. They usually are; if not finished at the time of original construction, they often are finished at a later date. Thus there is a total area of 2958 sq. ft.

There will be a 12-kW range and a 6.9-kW electric dryer. (For those who are not familiar with kW, 1 kW is 1000 watts, and 1 watt is 1 volt times 1 ampere.)

General Lighting:

2958 sq. ft. @ 3 watts per sq. ft.	8874	watts
Minimum of 2 small appliance circuits @		
1500 watts per circuit	3000	watts
Minimum of 1 laundry circuit @		
1500 watts	1500	watts
Total general lighting load	13,374	watts

From **Table 220-4(a)**:

3000 watts @ 100%	3000	watts
13,374 watts—3000 watts = 10,374		
watts @ 35%	3631	watts
	6631	watts
12 kW range (see **Table 220-19**)	8000	watts
6.9 kW dryer (no demand factor) or 100%	6900	watts
	21,531	watts

This 21,531 watts next has to be broken down into amperes per phase. A single-phase, 3-wire, 115/230-volt service will be used. Watts divided by volts equals amperes, and since this is single-phase 3-wire, the loads should be divided so that the amperage in each phase conductor is balanced

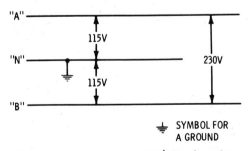

Fig. 7. A single-phase, 3-wire, 115/230-volt service.

insofar as possible. Fig. 7 illustrates the voltages present in a service of this type, and the two phases are designated by the letters A and B, and the neutral or grounded conductor by the letter N. Dividing 21,531 watts by 230 volts equals 93 amperes.

230-41. Size and Rating.

(b) **Ungrounded Conductors.** *Ungrounded conductors shall not be smaller than:*

(1) *100 ampere, 3-wire, for a one-family dwelling with six or more 2-wire branch circuits.*

> It is recommended that a minimum of 100 amperes 3-wire service be provided for all individual residences.

Please note the reference to *six 2-wire branch circuits.* This will be important and discussed later.

In Calculation No. 1 it was determined there was 93 amperes per phase. This is within the 100 amperes mentioned in **Section 230-41.** It would be wise to make the service entrance larger than the 100 amperes so that the service size would not have to be enlarged in the near future.

Some addition to the original calculation, though not specifically required, is also suggested. There was no garage figured, nor any lights or receptacles for porches or a patio. General-purpose outlets are calculated at 1½ amperes per outlet, and the fan motor on the furnace (if one is present) should be included.

Phase A	Phase B
93.0 amps.	93.0 amps.
5.2 amps. for fan	4.5 amps. for 3 outlets in garage
	3.0 amps. for outlets on porch (2)
98.2 amperes	100.5 amperes

The calculations just made will now be broken down into number and sizes of branch circuits.

General lighting was 2958 sq. ft. @ 3 watts per sq. ft., or 8874 watts. These will be 115-volt circuits, so 8874 divided by 115 volts equals 77.2 amperes. This would consist of a minimum of six 15-ampere circuits or four 20-ampere circuits for general lighting including receptacles for general purpose. These are the minimum number required, but consider adding a few more. There are also a minimum of two 20-ampere small appliance circuits (as explained in Chapter 2) and one 20-ampere laundry circuit.

Using the demand factor in **Table 220-19** for the 12 kW range, 8 kW was indicated. So, 8000 watts divided by 230 volts equals 35 amperes. Referring to **Table 310-16** and **310-18** of the NEC shows that this will require a minimum of No. 8 copper, or No. 6 aluminum in 60°C wire or No. 8 aluminum in 75°C wire. It is recommended that wire no smaller than No. 6 copper, or the equivalent in aluminum, be used.

The dryer was rated at 6900 watts, so 6900 divided by 230 volts equals 30 amps. No circuit should be loaded 100%. Since 80% loading is a safe figure, and since the loads of electric dryers have been increasing, use No. 8 copper or the equivalent in aluminum, and a 40-ampere circuit. Notice that 230 volts was used on the range and dryer calculations because both are 115/230-volt appliances.

The panel and service equipment will be covered in a later chapter covering services. Wire sizes and ampacity (ampere carrying capacity) will also be covered in a later chapter.

CALCULATION NO. 2

The optional calculation (**Table 220-30**), as shown in Chapter 2, will generally be used for homes that have considerable electrical equipment in them. However, the first example will be calculated on the Optional Basis and then the figures will be checked. There was neither electric heating or air-cooling, so:

General Lighting:

2958 sq. ft. @ 3 watts per sq. ft...........................8874 watts
2-20 amp. small appliance circuits @
    1500 watts each..3000 watts
1-20 amp. laundry circuit @ 1500 watts ....1500 watts
1-12 kW range (nameplate rating)..................12,000 watts
1.6.9 kW dryer (nameplate rating)....................6900 watts

                                                    32,274 watts

1st 10 kW of all other load at 100%................10,000 watts
Remainder of all other load @ 40%
    (21,274 watts) ....8910 watts

                                                    18,910 watts

Dividing 18,910 watts by 230 volts equal 83 amperes. Using the first calculation, we obtained 93 amperes, and using the optional calculation, we obtain 83 amperes.

CALCULATION NO. 3

Optional Calculation—**Table 220-30.**

This will be a 2,000 sq. ft. dwelling, exclusive of garage and porches, but including the basement. There will be a 1.2 kW dishwasher, a 1-kW disposal, 10 kW of space heating installed in 6 rooms (not a central heating plant) and a 6-ampere, 230-volt air conditioner (1380 watts divided by 1000 equals 1.38 kW). Note that the space heating wattage is larger than the air conditioner wattage, so the air cooling unit will not appear in the following calculations:

2000 sq. ft. @ 3 watts per sq. ft. .... 6000 watts
2-20 ampere small appliance circuits @
  1500 watts .... 3000 watts
1-20 ampere laundry circuit @ 1500 watts .. 1500 watts
1-12 kW range @ 100% ....12,000 watts
1-5 kW water heater @ 100% .... 5000 watts
1-1.2 kW wishwasher @ 100% .... 1200 watts
1-1.2 kW disposal @ 100% ....1000 watts
10 kW of space heating, divided
  in 6 rooms ....10,000 watts
                                                        39,700 watts

1st 10 kW @ 100% ....10,000 watts
29,700 watts @ 40% ....11,880 watts
                                                        21,880 watts

Then, 21,880 watts divided by 230 volts equals 96 amperes. So this installation will require a minimum service of 100 amperes.

CALCULATION NO. 4

The same figures used in Calculation No. 3 will be used here except for 8 kW of central electrical heating, 5 kW of air cooling, 2-4 kW wall-mounted ovens, and 1-6.5 kW counter-mounted cooking top.

As in Calculation No. 3, this will be a 2000 sq. ft. dwelling, including basement but excluding garage and porches. There will be 2-4 kW wall-mounted ovens, 1-6.5 kW counter-mounted cooking top, 8 kW of central heating, 5 kW of air cooling, 1-5 kW water heater, 1-1.2 kW dishwasher, and 1 kW disposal. The air cooling is larger in Calculation No. 3, and the heating is a central plant. Referring back to Chapter 2 on Optional Calculations, the air cooling can be ignored since it is smaller than the space heating load, and 100% of the central heating load is used since central heating does not have the diversity which 6 rooms on separate thermostats has. Also, there are now two ovens and one cooking top instead of a 12 kW range.

2000 sq. ft. @ 3 watts per sq. ft.	6000 watts
2-20 amp. small appliances circuits @ 1500 watts	3000 watts
1-20 amp. laundry circuit @ 1500 watts	1500 watts
Central space heating	8000 watts
Air cooling	None
2-4 kW wall-mounted ovens	8000 watts
1-6.5 kW counter-mounted cooking top	6500 watts
1-1.2 kW dishwasher	1200 watts
1-1 kW disposal	1000 watts
1-5 kW water heater	5000 watts
	40,200 watts

The 8 kW of central heating is subtracted before figuring the percentage and then added back later	−8000 watts
	32,200 watts
First 10 kW of the 32,200 watts @ 100%	10,000 watts
Remainder of 32,200 watts, or 22,200 @ 40%	8880 watts
Central heating @ 100%	8000 watts
	26,880 watts

Dividing 26,880 watts by 230 volts equals 116.8 amperes. This will require a service larger than 110 amperes. **Section 240-6** lists the next size breaker or fuse (standard size) as 125 amperes. It is suggested that at least a 125-ampere service and a 125-ampere breaker be used.

CALCULATION NO. 5

Table in Section 220-31

Load in kW or kVA	Percent of Load
First 8 kW of load at	100%
Remainder of load at	40%

Load calculation shall include lighting at 3 watts per square foot; 1500 watts for each 20 ampere appliance circuit, range or wall-mounted oven and counter-mounted cooking unit, and other appliances that are permanently connected or fastened in place, at nameplate rating.

If air conditioning equipment or electric space heating equipment is to be installed the following formula shall be applied to determine if the existing service is of sufficient size.

Air conditioning equipment*	**100%**
Central electric space heating*	**100%**
Less than four separately controlled space heating units*	**100%**
First 8 kW of all other load	**100%**
Remainder of all other load	**40%**

Other loads shall include:

1500 watts for each 20-ampere appliance circuit:
Lighting and portable appliances at 3 watts per sq. ft.
Household range or wall-mounted oven and counter-mounted cooking unit.
All other fixed appliances including four or more separately controlled space heating units, at nameplate rating.

*Use larger connected load of air conditioning and space heating, but not both.

This will cover an existing 115/230-volt or 120/208-volt, 3-wire, 60-ampere service to which an additional load is to be added. In this particular calculation, the present loads will be ignored and the following figures will be used to bring the wiring up to the 1981 NEC standards.

There are 1500 sq. ft. of finished residential area, a 12 kW range and a 3 kW air cooler.

1500 sq. ft. @ 3 watts per sq. ft. .... 4500 watts
2-20 ampere small appliance circuits @
   1500 watts .... 3000 watts

1-20 ampere laundry circuit @ 1500 watts   1500 watts
1-12 kW range .... 12000 watts
1-3 kW air cooler .... 3000 watts
                                                          ─────────────
                                                          24,000 watts

It is now necessary to use 100% of the air cooling load, which will leave 21,000 watts of other load;

Air cooling .... 3000 watts
1st 8 kW of other load .... 8000 watts
Balance 21,000 —8000 watts =
    13,000 watts @ 40% .... 5200 watts
                                                          ─────────────
                                                          16,200 watts

Dividing 16,200 watts by 230 volts equals 71 amperes. In this case, the 60-ampere service will have to be increased in ampacity, so refer back to **Table 220-11** or **220-30** and recalculate the service size. In doing so, it will be necessary to increase the service to the 100-ampere minimum.

MOTOR LOADS

See Chapter 2 and **Section 220-14.**

Where motor loads, air-cooling motors, furnace motors, pump motors, disposal motors, etc., are used, 125% of the full-load current rating of the largest motor is taken, plus 100% of the full-load current rating of the balance of the smaller motors, all added to the load for figuring the feeders and services. On branch circuits, use 125% of the full-load current rating of the motor or motors involved to calculate the conductor size for the branch circuit. This is assuming only one motor to a branch circuit. If there are two or more motors, take 125% of the largest full-load current rating and add 100% of the full-load current rating of other motors on the same branch circuit.

If the motor nameplate is in amperes instead of horse-power, use the full-load nameplate rating, but if the rating is in horsepower, refer to **Table 430-148** of the NEC to arrive at the amperes. In doing this, make note of whether the motors are 115 or 230 volt. The 230-volt motors will be added to both phase legs (A and B). If they are 115-volt motors, the load will be on one phase leg only and it will be necessary to balance the load on both phase legs as nearly as possible. Motor loads, except air-cooling and furnace motors, do not actually appear too frequently in residences.

It is recommended that reference be made to the Examples in Chapter 9 of NFPA 70 (NEC) or NFPA 70A where additional examples of calculations will be found to further check the method used in calculating.

CHAPTER 4

Farm Buildings

Information on farm buildings appears in **Section 220-40** and **220-41** and on farm services in **Section 220-40** of NFPA 70 (NEC). They do not appear in the ONE AND TWO-FAMILY DWELLING EDITION, NFPA 70A.

Calculations for farm buildings appeared for the first time in the 1965 NEC and take into consideration the demand factors for farm buildings and services. There are basically two types of service installations—the Main Service that goes to the dwelling and from which the farm buildings are served (see Fig. 8) and the Farm Service Pole from

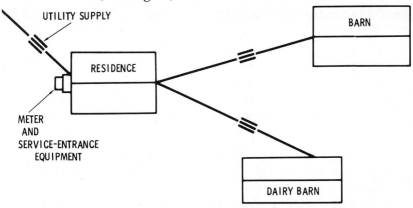

Fig. 8. A typical service installation for farm buildings in which the main service connects to the dwelling.

which the separate buildings, including the dwelling, are served by service drops and/or service laterals. The metering for this latter type is on the farm service pole (see Fig. 9). If wiring the dwelling, the loads in the other buildings must be considered in figuring the service equipment if the meter is at the dwelling. If a pole is used as the metering point, then the sizing of the conductors to the meter must be figured.

Probably the best way to present this is to quote the NEC and then follow with a calculation as an example.

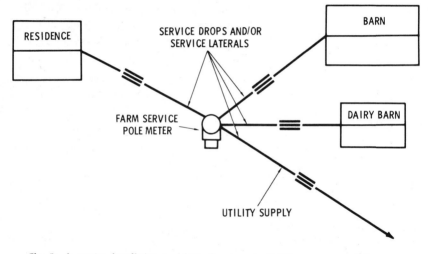

Fig. 9. A service installation in which the separate buildings are served from a service pole.

Section 220-41. Farm Loads—Total. *The total load of the farm for service-entrance conductors and service equipment shall be computed in accordance with the farm dwelling load and demand factors specified in* **Table 220-41.** *Where there is equipment in two or more farm equipment buildings or for loads having the same function, such loads shall be computed in accordance with* **Table 220-40** *and may be* *combined as a single load in* **Table 220-41** *for computing the total load.*

See **Section 230-21** *for overhead conductors from a pole to a building or other structure.*

Table 220-40. Method for Computing Farm Loads For Other Than Dwelling Unit

Ampere Load at 230 Volts	Demand Factor Percent
Loads expected to operate without diversity, but not less than 125 Percent full-load current of the largest motor and not less than the first 60 amperes of load 100	
Next 60 amperes of all other loads 50	
Remainder of all other loads 25	

Table 220-41. Method of Computing Total Farm Load

Individual Loads Computed in Accordance with Table 220-40	Demand Factor Percent
Largest load 100	
Second largest load 75	
Third largest load 65	
Remaining loads 50	

To this total load, add the load of the farm dwelling computed in accordance with **Part B** *and* **C** *of this Article.*

Note: Computation of dwelling loads was covered earlier.

EXAMPLE OF A FARM CALCULATION

This example excludes the farm residence.

Load No. 1 (Feed Grinder and Auger)

5-HP, single-phase, 230-volt motor	28 amperes
1-HP, single-phase, 230-volt motor	8 amperes
Load No. 1 Total	36 amperes

Load No. 2 (Milk Barn)
Computed Load

Lighting	2000 watts
Water heater	2500 watts
Total	4500 watts

4500 watts divided by 230 volts	20 amperes
2-HP milker	12 amperes
1-HP cooler	8 amperes
Air conditioner	15 amperes
Load No. 2 Total	55 amperes

Load No. 3 (Chicken House)

Brooder	3000 watts
Lighting	450 watts
	3450 watts

Load No. 3 total (3450 watts divided by 230 volts)	15 amperes

In these computations, the largest motor is 5 HP rated at 28 amperes. This will have to be increased by 25% of full-load current, so it will be 35 amperes.
The computed load is as follows:

Load No. 1	43 amperes
Load No. 2	55 amperes
Load No. 3	15 amperes
Total	113 amperes
First 60 amperes at 100%	60 amperes
Next 60 (or less) amperes @ 50%	27 amperes
Remainder @ 25%	(0) amperes
Computed load using **Table 220-11**	87 amperes

Thus, the service to supply the farm buildings will not be the 113 amperes, but will be 87 amperes.

42

The next example is where there might be a farm service pole and service drops to all of the loads including the dwelling.

The single-family dwelling has a floor area of 1500 sq. ft. exclusive of an unfinished attic and porches. It has a 12-KW range.

Residence computed load (see **Sections 220-11** thru **220-19**. Also refer back to Chapter 3 which covered the computations for dwelling occupancies.

General Lighting Load:

1500 sq. ft. @ 3 watts per sq. ft.	4500 watts
Small appliance circuits	3000 watts
Laundry circuit	1500 watts
Total without range	9000 watts
3000 watts at 100%	3000 watts
9000 watts—3000 watts at 35%	2100 watts
Net computed load without range	5100 watts
Range load (see **Table 220-19**)	8000 watts
Net computed load with range	13,100 watts

Load No. 1 total for 115/230-volt, 3-wire system feeders:

13,100 divided by 230 volts	57 amperes

Load No. 2 (Feed Grinder and Auger)

5-HP single-phase, 230-volt motor	28 amperes
1-HP single-phase, 230-volt motor	8 amperes
Load No. 2 Total	36 amperes

Load No. 3 (Milk Barn)

Lighting	2000 watts
Water heater	2500 watts
Total	4500 watts

4500 watts divided by 230 volts	20	amperes
2-HP milker	12	amperes
1-HP cooler	8	amperes
Air conditioner	15	amperes
Load No. 3 Total	55	amperes

Load No. 4 (Chicken House)

Brooder	3000	watts
Lighting	450	watts
Total	3,450	watts

Load No. 4 (3450 watts divided by 230 volts) .... 15 amperes

From **Table 220-41:**

This load includes the dwelling and is served from a farm service pole and service drops to all buildings.

Largest demand (residence)		
57 amperes @ 100%	57	amperes
Second largest (demand Load No. 3)		
55 amperes @ 75%	42	amperes
Third largest demand (Load No. 2) with 25% of the 28 amps. for the 5-HP motor added—28 + 7 = 35, or a total of 43 amperes @ 65%	28	amperes
Balance of the demand (Load No. 4)		
15 amperes @ 50%	8	amperes
	135	amperes

The total connected load for the farm would be 170 amperes, but using the demand factors in **Section 220-40** and **41**, a service to handle 135 amperes would be installed. Bear in mind that this would not allow any expansion for future demand.

CHAPTER 5

Services (General)

The services, service entrance, service equipment, and the grounding of services are all extremely important subjects. In Chapter 2, definitions of the various parts of services were given. Refer to these to refresh your memory as only the terminology will be used in this chapter.

Most inspectors will start any inspection at the service point. It is here that clues will be found as to what to expect in other portions of the wiring system. It is the service and service equipment, plus the grounding, that is the watchdog in the protection of the rest of the wiring system. Here will be found the protection against overloads and faults, and grounding for the protection of the system from shocks, lightning, breakdown of the supply transformer windings, etc.

First, consider the ampacity of the service in its entirety. The calculations involved for services were covered in Chapter 2, while examples were given in Chapter 3. In addition, Chapter 9 of the NEC has additional samples.

Section 230-23 of the NEC (**Minimum Size of Service Drop Conductors**) states: *Conductors shall have sufficient ampacity to carry the load. They shall have adequate mechanical strength and shall not be smaller that No. 8 copper or No. 6 aluminum or copper-clad aluminum.* An exception is included, but is not pertinent to this discussion.

Section 230-31 of the NEC (**Size of Underground Service conductors**) states: *Conductors shall have sufficient ampacity to carry the load. They shall not be smaller than No. 8 copper or No. 6 aluminum or copper-clad aluminum.* Again an exception is listed that is not pertinent to this discussion.

In **Section 230-41** is found a pertinent exception: Exception No. 1. For single-family residences with an initial load of 10 kW or more computed in accordance with **Article 220,** or if the initial installation has more than six 2-wire branch circuits, the service-entrance conductors shall have an ampacity of not less than 100 amperes 3-wire.

These Sections, especially **Section 230-41,** tell us a great deal. In broad terms they indicate that a 100-ampere service is the minimum permitted for a single-family residence. Note the word **minimum.** Many larger services are required, as in large residences and in electrically heated residences. Also remember that, when considering the NEC, minimum requirements are always discussed. I am often told that the utilities usually find residential demands much less than Code requirements. This no doubt is true, but just what is our responsibility as inspectors and yours as wiremen, and what is the responsibility of the NEC? Wiring is not installed just for today, but for anticipated future needs.

An interesting case was recently encountered where the home owner requested a 150-ampere service. By calculations, however, a 100-ampere service would have sufficed. The utility furnished overcurrent protection of only 70 amperes on the meter pole. The wireman was extremely concerned as to why only 70-ampere protection was provided for the 150-ampere service. He really needn't have been concerned as the utility owned and maintained all the equipment on the meter pole, and under **Section 90-2(b)** of the NEC, this part was exempted from the Code. Why the concern—the customer ordered a 150-ampere service and got it. If the 70-

46

ampere overcurrent protection on the meter pole was inadequate, the utility would supply whatever might be needed at their expense. In the meantime, the customer's home was wired adequately for many years to come.

Be sure that the service size meets the requirements of **Article 220** with a 100-ampere ampacity (or larger if the calculations require a larger service).

For single copper conductor service drops in free air, use **Table 310-17**. For aluminium conductors, use Table **310-19**. Note that the ampacities in these two tables are greater than those in **Tables 310-16** and **310-18**. This is because of the greater heat dissipation in free air.

Note 3, accompanying **Tables 310-16** through **310-19**, gives alternate ratings for service-entrance conductors.

3. Three-Wire, Single-Phase Residential Service. *In dwelling units, conductors, as listed below, shall be permitted to be utilized as three-wire, single-phase, service entrance conductors and the three-wire, single-phase feeder that carries the total current supplied by that service.*

Conductor Types and Sizes
RH-RHH-RHW-THW-THWN-THHN-XHHW

Copper	Aluminum and Copper-Clad AL	Service Rating in Amperes
AWG	AWG	
4	2	100
3	1	110
2	1/0	125
1	2/0	150
1/0	3/0	175
2/0	4/0	200

SERVICE-ENTRANCE LOCATION

Where should the service entrance and service equipment be located? This is sometimes quite an involved problem, but not in all cases.

First of all, the serving utility is concerned as to which portion of the residence a service drop is to be installed. Service laterals do not seem to cause as much of a problem since they do not have to run in a direct line.

According to the NEC:

Section 230-72(c and d). Location. The disconnecting means shall be located at a readily accessible point nearest to the entrance of the conductors, either inside or outside the building or structure. Sufficient access and working space shall be provided about the disconnecting means.

Section 230-44. Conductors Considered Outside Building. Conductors placed under at least two inches of concrete beneath a building, or conductors within a building in conduit or duct and enclosed by concrete or brick not less than two inches thick shall be considered outside the building.

Consider all of the preceding items when locating the service-entrance and the service equipment.

Inspection authorities differ on the length of the service-entrance conductors to the service equipment. Check with the local authority having jurisdiction. Refer to **Section 230-72(c)**. This section states a great deal in very few words. What does the statement *it shall be readily accessible* mean? To me it indicates that the disconnecting means shall not be in a bedroom, bathroom, dish cupboard, etc., and I will not accept a disconnect in the basement unless there is a ground-level exit. The disconnecting means is an emergency item and I am certain that a basement would be the last place I would wish to go in the event of a fire. Bedrooms and bathrooms are private rooms, and the bathroom has grounded items in addition to the steam and moisture present which could create a hazard. The NEC also states that panelboards shall not be located *near com-*

bustibles. I interpret this to mean they cannot be located in clothes closets and dust-mop closets.

Take note of the statement, *point nearest the entrance of the conductors.* Here the inspector must use his powers of interpretation as granted in **Section 90-4.** To me this means not to exceed approximately 15 feet. Even 15 feet might be too far in some cases. I will often require overcurrent protection at the outer end. If in doubt, check with the authority having jurisdiction.

Sufficient access and working space shall be provided about the disconnecting means. Basically, this means that clearance must be provided so that an electrician does not have to lean across a washer, dryer, or other appliance in working on the equipment, or have to use a ladder or chair, or have any obstruction in his way when working on the panel.

Section 230-90 states that the service-entrance conductors shall be the same ampacity as the main or larger. **Section 384-16(a), Exception 2** states that a split-bus panel might be used in residences.

INSTALLATION OF SERVICE DROPS

If a service drop is used, check the point of attachment. Make sure that all requirements of **Section 230-24** in the NEC are met.

Section 230-24. Clearance of Service Drops. *Service-drop conductors shall not be readily accessible and shall comply with* (a) *through* (c) *below for services not over 600 volts, nominal.*

(a) **Above Roofs.** *Conductors shall have a vertical clearance of not less than 8 feet (2.44 m) from all points of roofs*

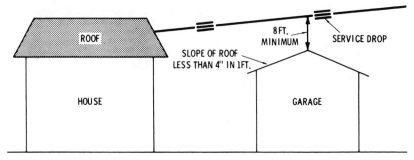

Fig. 10. Clearance of conductors passing over roofs must conform to Code rulings.

above which they pass. (See Fig. 10.)

Exception No. 1. Where the voltage between conductors does not exceed 300 and the roof has a slope of not less

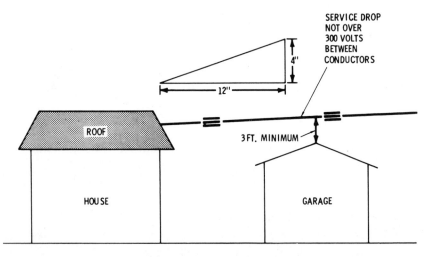

Fig. 11. Clearance of conductors passing over roofs are governed by voltage and by the roof slope.

than 4 inches in 12 inches, the clearance may be not less than 3 feet (914 m). (See Fig. 11.)

Exception No. 2. Where the voltage between conductors does not exceed 300, a reduction in clearance above only the overhanging portion of the roof to not less than 18 inches (457 mm) shall be permitted if (1) not more than 4 feet (1.22 m) of the service-drop conductors passes above the roof overhang, and (2) they are terminated at a through-the-roof raceway or approved support. (See Fig. 12.)

See Section 230-28 *for mast supports.*

(b) **Clearance from Ground.** This is covered in **Section 230-24(b)** of the NEC. The Section will not be copied here but is shown by Fig. 13.

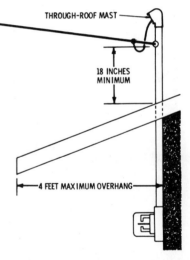

Fig. 12. Conductors passing over a portion of a roof and terminating at a through-the-roof service raceway have different clearance requirements.

(c) **Clearance From Building Openings.** This is covered by **Section 230-24(c)** of the NEC. The intent is shown in Fig. 14.

If a mast is required for clearances, refer to **Section 230-28** of the NEC. This mast shall be strong enough to support the service drop in sleet storms, high winds, etc., or

51

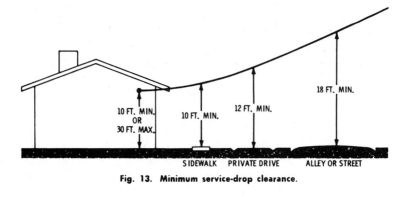

Fig. 13. Minimum service-drop clearance.

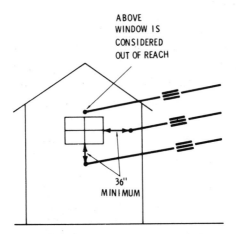

Fig. 14. Minimum service-drop clearance around building openings.

will require guying. In my inspection area, no smaller than 2-inch rigid galvanized conduit is accepted, and guying as in the judgment of the inspector is required. See Fig. 15.

INSTALLATION OF SERVICE LATERALS

Service laterals may be direct-burial conductors or cable, or may be in approved raceways. Refer to **Section 230-49.**

230-30. Insulation. Service lateral conductors shall be insulated for the applied voltage.

Exception: A grounded conductor may be:

(a) *Bare copper in a raceway.*

(b) *Bare copper for direct burial where bare copper is judged to be suitable for the soil conditions.*

(c) *Bare copper for direct burial without regard to soil conditions where part of an approved cable assembly with a moisture- and fungus-resistant outer covering.*

(d) Aluminum or copper-clad aluminum without individual insulation or covering used in a raceway or for direct burial when a part of a cable assembly approved for the purpose and having a moisture- and fungus-resistant outer covering.

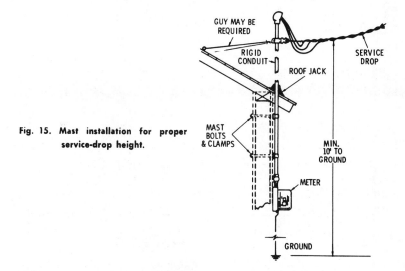

Fig. 15. Mast installation for proper service-drop height.

Take note of the bare copper for direct burial, but also take note of the fact that the soil conditions must be judged suitable and this judging is up to the authority having jurisdiction, as provided in **Section 90-4**.

Approved raceways may be used. This could be duct, nonmetallic conduit, or rigid galvanized conduit (provided the galvanizing is approved by the authority having jurisdiction). Most authorities will require additional corrosion protection over the galvanizing, as outlined in **Section 346-1**. UNDERWRITER'S LABORATORIES have conducted tests on the corrosion of galvanized conduits, and have established standards based on the ohms-per-centimeter resistivity. Since this test requires special testing equipment, most authorities will insist on additional corrosion protection. The service laterals may be installed in these approved raceways, but the insulation of the conductors shall meet the requirements of

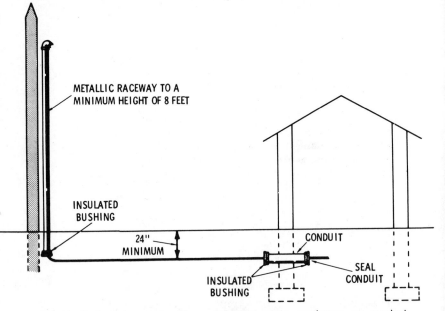

METALLIC RACEWAY TO A
MINIMUM HEIGHT OF 8 FEET

INSULATED
BUSHING

24"
MINIMUM

CONDUIT

INSULATED
BUSHING

SEAL
CONDUIT

Fig. 16. **Mechanical protection for underground service conductors are required where they enter a building or leave the ground to go up a pole.**

Section 310-7 (**Wet Locations**) Types RUW, RHW, TW, THW, THWN, XHHW, lead-covered, etc., may be installed.

For direct burial, the conductors shall be Type USE buried a minimum of 24 inches deep. The authority enforcing the Code may require supplemental mechanical protection, such as a covering board, concrete pad, or raceway. In rocky soil, and more especially where frost is prevalent, the inspection authority will usually require a fine sand bed with a fine sand covering under and over the conductors. Rocks subjected to frost heave will cause damage to the insulation.

Mechanical protection is also required where entering the building or when leaving the ground to go up a pole. See Fig. 16. and **Section 230-50.**

Raceway sealing is required at the building to prevent the entrance of moisture and/or gases. Duct seal may be used. Service-lateral conductors shall be without splice, but where they enter the building they cease to be service-lateral conductors and become service-entrance conductors, so a splice is permitted at that point. See Fig. 17.

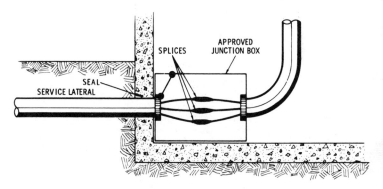

Fig. 17. A splice in an underground system.

SERVICE ENTRANCES AND EQUIPMENT

A portion of this subject was covered earlier in the chapter as it is practically impossible to separate service drops from service laterals. Figs. 12 and 15 showed service masts and the conductors from the tap to the service drop, on down to the meter. From this point to the service equipment are the service-entrance conductors.

Section 230-46 tells us that service-entrance conductors shall be without splice, but it is known that these conductors will, by necessity, have to be broken in the meter box, a subject covered by Exception No. 1 which permits clamped or bolted connections in meter housings.

Fig. 18 shows service entrances other than a mast type. Where the proper height can be obtained without a mast,

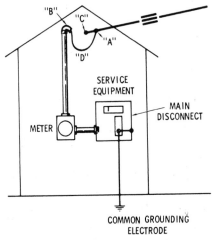

Fig. 18. A typical service installation.

as shown in Fig. 13, such an installation may be used. This might be service-entrance cable (**Article 338**), rigid conduit (**Article 346**), or electrical metallic tubing (**Article 348**), provided watertight fittings are used.

The service-entrance conductors extend from point A in Fig. 18 to the line side of the main disconnecting means in the service equipment. Notice that the service-entrance head (B) is higher than the point of attachment of the service drop. This is covered in **Section 230-54 (Connections at Service Head)**. If it is not possible to install the service head higher than the point of attachment, there are other alternatives. However, the basis of all this is to keep water from running into the service raceway and equipment. Notice the drip loop (D) for this purpose. I often suggest a very small notch in the insulation at the bottom of the drip loop. This will break the syphoning effect. The newer plastic insulations do not adhere to the conductors as the old type rubber insulation did, and this creates syphon effects.

Some utilities will require more than the minimum 10 feet for the point of attachment allowed by the Code. For

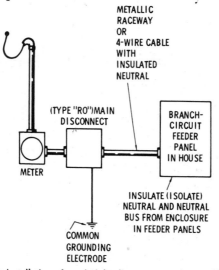

Fig. 19. **Proper installation of a raintight disconnect on the outside of a house.**

instance, one utility in my area requires a minimum of 12 feet for the point of attachment.

How high should the meter housing be installed? This depends upon the utility requirements, but in most cases it will not be less than 5 feet nor more than 6 feet high.

Basically, service equipment is defined as the circuit breakers, fuses, or switches and their accessories to be used as a main disconnect to the residence. In the early portion of this book, a farm service pole was shown. If there is an overcurrent device mounted with the meter on this pole, this device is purely an overcurrent device and cannot be termed the main disconnect. **Section 230-72(c)** states that it shall be on the inside or outside of the building or structure at the closest point of entrance of the conductors. In the majority of cases, the main disconnect for a residence will be in an enclosure along with the branch circuit breakers or fuses. Remember that, in **Section 384-16**, split-bus panels are permitted to be used in residences.

A raintight main disconnect (RO) may be installed on the outside of the house to serve as the main disconnect, with a feeder circuit from this main to a branch-circuit panel in the house. If this is the case, the conductors from the main to the branch-circuit panel are feeders and shall have an ampacity equal to or larger than the main disconnect. Also, an equipment ground conductor from the main to the branch circuit is required, with the neutral insulated from the enclosure of the branch-circuit panel. This equipment grounding conductor may be a metal raceway (conduit or EMT), or if the feeder is a cable, this cable shall have an insulated neutral, two phase conductors, and an equipment grounding conductor. See Fig. 19.

The branch-circuit panel buses shall have an ampacity equal to or greater than the rating of the main disconnect breaker, switch, or fuses. If the branch-circuit buses of a feeder panel are not of this capacity, then a main breaker shall be installed to protect the buses. The only place you might run into this in a residence is where there are two or more feeder panels. See **Section 384-16.**

There is nothing in the Code prohibiting the installation of a branch-circuit panel with a main disconnect or a split-bus panel on the outside of the residence, provided a rain-tight enclosure is used (RO). See Figs. 19 and 20.

Recall that **Section 230-41(b)(1), Exception No. 1** states: if the initial installation has more than six 2-wire circuits, the service-entrance conductors shall have an ampacity of not less than 100 amperes 3-wire.

It should be mentioned here that white-colored conductors shall never be used for phase conductors—neither may they be marked with some other color. They are to be used

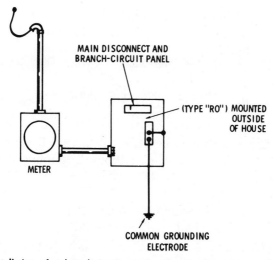

Fig. 20. Installation of a branch-circuit panel and main disconnect on the outside of a house.

strictly as the grounded conductor (neutral). This applies to natural gray color as well. However, **Section 200-6** permits insulated conductors larger than No. 6, other than white or natural gray, to be used as the grounded conductor (neutral), provided they are plainly identified at the terminations with white at the time of installation.

There are some utilities that require one meter for the residence proper and one for the water heater, with an off-peak time clock and a different rate structure. While **Section 373-8** is not applicable to utilities or to meter housings owned by the utility, you will no doubt find that they will require the same type of installation as the Code requires. See Fig. 21.

373-8. Enclosures for Switches or Overcurrent Devices. *Enclosures for switches or overcurrent devices shall not be used as junction boxes, auxiliary gutters or raceways for conductors feeding through or tapping off to other switches or overcurrent devices.*

Notice that nothing is mentioned about meter enclosures, but you will no doubt be required to install housings as shown in Fig. 21.

GROUNDING OF SERVICES

Grounding is covered by **Article 250** of the NEC, and in covering this portion of services, we should become familiar with the terminology and definitions pertaining to grounds. See Definitions in **Article 100.**

> **Ground:** *A conducting connection, whether intentional or accidental, between an electrical circuit or equipment and earth, or to some conducting body which serves in place of the earth.*

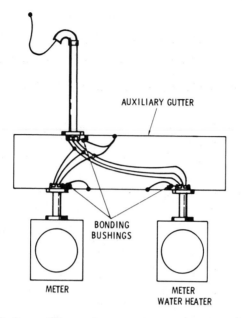

AUXILIARY GUTTER

BONDING
BUSHINGS

METER

METER
WATER HEATER

Fig. 21. Some utilities require a separate meter for a water heater.

Grounded Conductor: *A system or circuit conductor which is intentionally grounded.* This would be what we term the neutral. Residences are wired with 115/230 volts or 120/208 volts. See Fig. 22.

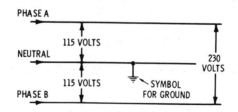

PHASE A

115 VOLTS

NEUTRAL

230
VOLTS

115 VOLTS

PHASE B

SYMBOL
FOR GROUND

Fig. 22. A 115/230-volt, single-phase, 3-wire system.

Grounding Conductor: *A conductor used to connect equipment or the grounded circuit of a wiring system to a grounding electrode or electrodes.*

61

Grounding Electrode Conductor. *The conductor used to connect the grounding electrode to the equipment grounding conductor and/or to the grounded conductor of the circuit at the service equipment or at the source of a separately derived system.*

Grounding of services, as previously stated, is very important and must be done properly. Different situations will be covered to give examples of various problems which might be encountered.

In order to summarize, **Section 250-51 (Effective Grounding)** will be broken down here.

The path to ground from circuits, equipment, and conductor enclosures shall

(1) be permanent and continuous and

(2) shall have ample carrying capacity to conduct safely any currents liable to be imposed on it, and

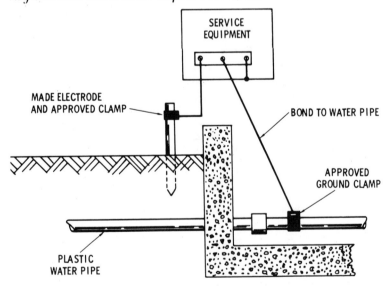

Fig. 23. Approved method of grounding a wiring system.

(3) shall have impedance sufficiently low to limit the potential above ground and to facilitate the operation of the overcurrent devices in the circuit.

Part (3) is often ignored, probably because of the word impedance. Impedance is AC resistance. **Section 110-10** in the NEC emphasizes its importance. Fault currents available are constantly increasing in amplitude. This refers to phase-to-phase faults, phase-to-ground faults, and to both bolted and arcing faults. Part (3) points out to be ever mindful of impedances in connection with grounding, and to keep these impedances as low as possible.

Section 250-23. Grounding Connections for Alternating-Current Systems. The grounding conductor (common main grounding connector) of a wiring system shall also be used to ground equipment, conduit, and the supply side of the disconnecting means. This is done so that when the disconnecting means is opened, the grounded conductor will not be opened, which would interrupt the grounding on the system. On a service of high capacity, it is recommended that the grounding conductor be connected within the service-entrance equipment. Personally, I find it should always be connected within the service equipment regardless of ca-

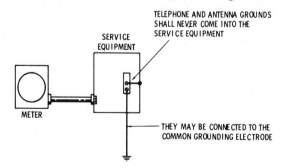

Fig. 24. Proper method of grounding telephones and antennas.

pacity. This is my preference in order that service work might be more readily performed. If the grounding connection is within the meter housing, it is not always available to the electrician who is doing the service work. See Fig. 23.

There have been occasions where the telephone grounding conductor or an antenna ground was run into the service equipment. This should never be permitted. It is proper to tie them to the same common grounding electrode, but they should never be in the same enclosure. See Fig. 24.

GROUNDING ELECTRODE

This portion of the Code is the basis and requirements for a properly installed grounding system. Great care should be taken to thoroughly understand the contents of this portion of the book and to adhere to the installation of the grounding system as required by the NEC. **Article 250** is constantly changing as better grounding methods become known.

It should be stated at this point that if the common grounding conductor is connected into the meter housing, or for that matter into the service equipment, be careful to note that there shall **not** be an aluminum service-entrance conductor and a common grounding conductor under the same lug. The restrictions put on the use of aluminum for a common grounding conductor rules it out, so copper must be used. If the copper conductor and the aluminum conductor are in the same connector, the copper will cause electrolysis on the aluminum conductor and make a high-resistance connection. See Fig. 25.

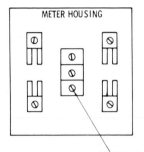

METER HOUSING

NEVER INSTALL ALUMINUM
AND COPPER CONDUCTORS
IN SAME LUG OR CONNECTOR

Fig. 25. Aluminum and copper conductors must not be fastened together with the same connector.

H. Grounding Electrodes

This part of **Article 250** is the basis for the proper construction of a grounded system. Great care should be taken to thoroughly understand the contents of this part, and to adhere to the installation as required by the Code.

250-81. Water Pipe. A buried metallic underground water-supply system shall always be used as the grounding electrode wherever there are 10 feet or more of buried pipe, including any well casing that is bonded to the system. If there is a chance that the piping system will be disconnected, or that an insulated coupling is or will be installed, or the possibility that nonmetallic water piping might be installed at a later date, the pipe electrode shall be supplemented by one or more made electrodes bonded to the piping. See Figs. 26, 27 and 28.

Section 250-81(a) now requires that a metal underground water pipe shall be supplemented by an additional electrode of a type specified in **Section 250-81** or in Section **250-83**. See Fig. 26.

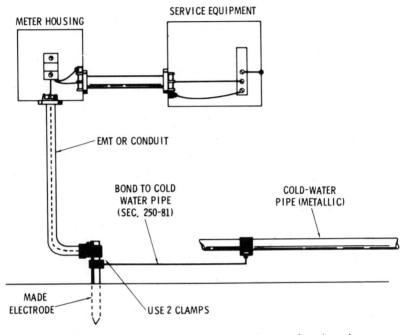

Fig. 26. Interior cold-water piping bonded to the grounding electrode.

A new addition was that an interior metallic cold-water piping system shall always be bonded to the one or more grounding electrodes. Note that the electrical wiring system should be adequately grounded without depending on the outside piping system. This means that supplementing the water piping system with made electrodes is advised. Additional safety may be gained by bonding the grounding electrode to the gas, sewer, and hot water piping, and to metallic air ducts within the building.

250-81. Grounding Electrode Systems. In addition to the underground metallic pipe system, it shall be supplemented by, or if the metallic underground metallic water pipe system is not available, the following may be used:

66

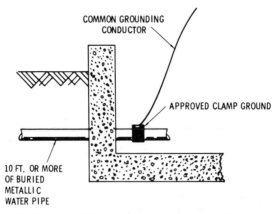

Fig. 27. Buried metallic water pipe used as a grounding electrode.

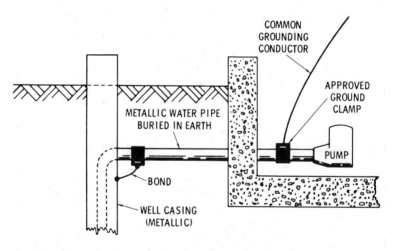

Fig. 28. Buried metallic water pipe bonded to a well casing and used as a grounding electrode.

(a) **Metal Frame of Building.** *The metal frames of buildings that are effectively grounded.*

(b) **Concrete Encased Electrodes.** *An electrode encased by at least 2 inches of concrete, located within and near the bottom of a concrete foundation or footing that is in direct* *contact with the earth, consisting of at least 20 feet of one or more steel reinforcing bars or rods of not less than ½ inch in diameter, or consisting of at least 20 feet of bare solid copper conductor not smaller than No. 4 AWG.*

(c) **Ground Ring.** *A ground ring encircling the building or structure in direct contact with the earth at a depth below earth surface not less than 2½ feet, consisting of at least 20 feet of bare copper conductor not smaller than No. 2 AWG.*

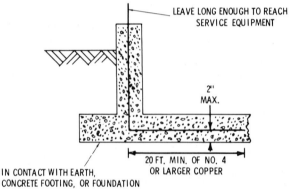

Fig. 29. Bare copper wire encased in concrete and used as a grounding electrode.

250-83. Made and Other Electrodes. *Where none of the electrodes specified in Section 250-81 is available, one or more of the electrodes specified in (a) through (d) below shall be used. Where practicable, made electrodes shall be embedded below permanent moisture level. Made electrodes shall be free from nonconductive coatings, such as paint or enamel. Where more than one electrode system is used (including those used for lightning rods), each electrode of one system shall not be less than 6 feet from any other electrode of another system.*

Two or more electrodes that are effectively bonded together are to be treated as a single electrode system in this sense.

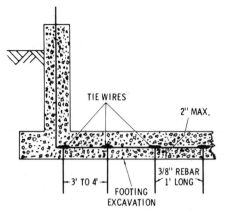

TIE WIRES

2" MAX.

3/8" REBAR
1' LONG

3' TO 4'

FOOTING
EXCAVATION

Fig. 30. Using rebars to improve the grounding effectiveness of a copper wire encased in a concrete footing or foundation.

(a) **Metal Underground Gas Piping System.** *An electrically continuous metal underground gas piping system that is uninterrupted with insulating sections or joints and without an outer nonconductive coating, and then only if acceptable to and expressly permitted by both the serving gas supplier and the authority having jurisdiction.*

(b) **Other Local Metal Underground Systems or Structures.** *Other local metal underground systems or structures, such as piping systems and underground tanks.*

(c) **Rod and Pipe Electrodes.** *Rod and pipe electrodes shall be not less than 8 feet in length and shall consist of the following materials, and shall be installed in the following manner:*

(1) *Electrodes of pipe or conduit shall not be smaller than ¾ inch trade size and, where of iron or steel, shall have the outer surface galvanized or otherwise metal-coated for corrosion protection.*

(2) *Electrodes of rods of steel or iron shall be at least ⅝ inch in diameter. Nonferrous rods or their equivalent shall be listed and be not less than ½-inch in diameter.*

69

(3) *The electrode shall be installed such that 8 feet (2.44 m) of length is in contact with the soil. It shall be driven to a depth of not less than 8 feet (2.44 m) except where rock bottom is encountered. The electrode shall be driven at an oblique angle not to exceed 45 degrees from the vertical or shall be buried in a trench that is at least 2½ feet (762 mm) deep. The upper end of the electrode shall be flush with or below ground level unless the above ground end and the grounding electrode conductor attachment are protected against physical damage as specified in* **Section 250-117.**

(d) **Plate Electrode.** *Each plate electrode shall expose not less than 2 square feet of surface to exterior soil. Electrodes of iron or steel plates shall be at least ¼ inch in thickness. Electrodes of nonferrous metal shall be at least 0.06 inch in thickness.*

Each electrode shall be separated by at least 6 feet from other electrodes used for signal circuits, radio, lightning rods, television, and other purposes. Although not in the Code, it is considered good practice to bond the electrodes together. In fact, many problems will be solved by doing this.

250-84. Resistance. Made electrodes shall have a resistance to ground of 25 ohms or less, wherever practicable. When the resistance is greater than 25 ohms, two or more electrodes may be connected in parallel or extended to a greater length. The Code cannot go into the mechanics of grounding, but good practice indicates that the electrode has a lower resistance when driven some distance from a foundation into undisturbed soil where the earth will put pressure on the driven electrode.

Water piping usually has a resistance of 3 ohms or less. Metal frames of buildings often make a good ground, especially where they contact rebar in the concrete, and usually have a resistance of less than 25 ohms. As was pointed out in **Section 250-82(a)**, the metal frame of a building (when effectively grounded) may be used as the ground. Local metallic water systems and well casings also make good grounds in most cases. One might wonder why metal frames are covered in dealing with residences, but there is a trend to new types of residential construction, and one of these uses metal studs.

Grounding, when made electrodes are used, can be greatly improved by the use of chemicals such as magnesium sulfate,

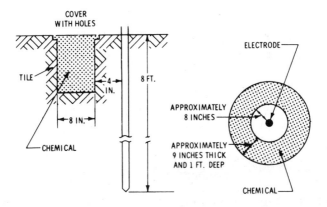

Fig. 31. Adding chemicals to the soil to lower its ground resistance.

copper sulfate, or rock salt. A doughnut-type hole may be dug around the ground rod into which the chemicals are put. Another method is to bury a tile close to the rod and fill the tile with the chemical. Rain and snow will disolve the chemicals and allow them to penetrate the soil, lowering its resistance. See Fig. 31.

The NEC recommends that the resistance of the grounds be tested periodically after installation. This is rarely done. In fact, it is practically never done even at the

71

time of installation, except by utility companies who realize the importance of an adequate ground. The testing of grounds is a mystery to many electricians. Never attempt to use a common ohmmeter for the testing–the readings obtained are apt to be most anything due to stray AC or DC currents in the soil or due to DC currents set up by electrolysis in the soil. There are many measuring devices on the market, such as the grounding MEGGER (trade mark of the James G. Biddle Co.), battery-operated ground testers that use vibrators to produce pulsating AC current, etc. In recent years, a transistor-type ground tester has appeared on the market.

An example of what might be expected by paralleling ground rods is as follows (these figures are general and should not be taken as the results in every case): Two rods in parallel, with a 5 ft. spacing between them, will reduce the resistance to about 65% of what one rod would be. Three rods parallel with the same spacing will reduce the resistance to about 42%, while four rods paralleled will reduce the resistance to about 30%.

J. Grounding Conductors

250-91. Material. The material for grounding conductors shall be as follows:

(a) **For System or Common Grounding Conductors.**
 (1) Copper or other corrosive-resistant materials.
 (2) Solid or stranded.
 (3) Insulated or bare.
 (4) Without splice or joints, except in the case of bus bars.
 (5) Electrical resistance per foot (linear) shall not exceed that of the allowable copper conductors that might be used for this purpose. Thus, if aluminum

(in cases where permissible) is used, the conductor will have to be larger than copper would be for the same purpose.

(b) For Conductor Enclosures and Equipment Only. The grounding conductor for equipment, conduit, and other metal raceways or conductor enclosures may be:

(1) Copper or other corrosive-resistant material.
(2) Bus bar, rigid conduit, steel pipe, EMT, or the armor of AC metal-clad cable. *Intermediate metal conduit.*
(3) Stranded or solid.
(4) Joints of conduit must be made up wrench tight, not plier tight.

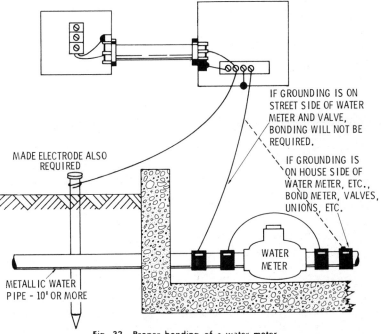

MADE ELECTRODE ALSO REQUIRED

IF GROUNDING IS ON STREET SIDE OF WATER METER AND VALVE, BONDING WILL NOT BE REQUIRED.

IF GROUNDING IS ON HOUSE SIDE OF WATER METER, ETC., BOND METER, VALVES, UNIONS, ETC.

WATER METER

METALLIC WATER PIPE - 10' OR MORE

Fig. 32. Proper bonding of a water meter.

It is well to state at this point that where there is 10 ft. or more of buried metallic water piping used as the common grounding electrode (**Section 250-81**), and where a water meter is present, it will be required to either connect the common grounding conductor on the street side of the water meter, or if connected on the house side, the water meter will have to be bonded. See **Section 250-80.** This will no doubt be interpreted to mean valves, unions, etc. See **Fig. 32.**

250-92. Installation. Grounding conductors shall be installed as follows:

(a) **System or Common Grounding Conductor.** No. 4 or larger conductors may be attached to the surface—knobs or insulators are not required. Mechanical protection will be

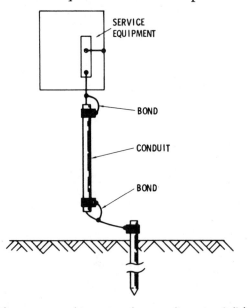

Fig. 33. Conduit or armor used to protect the grounding wire shall be bonded to the grounding electrode and to the service-entrance equipment.

required only where the conductor is subject to severe physical damage. No. 6 grounding conductors may be run on the surface of a building if protected from physical damage and rigidly stapled to the building structure. Grounding conductors smaller than No. 6 shall be in conduit, EMT, or armor. One might just as well forget No. 8 copper as a common grounding conductor unless it is run in a raceway.

The metallic enclosures for the grounding conductor shall be continuous from the cabinet to the grounding electrode and shall be attached at both ends by approved methods. See Fig. 33. If the conduit, etc., is used for protection purposes only, then the common grounding conductor shall be bonded to the metalic raceway at one or both ends as re-

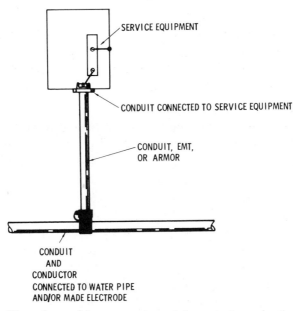

Fig. 34. Where the conduit or armor is used for protection only, the common grounding conductor shall be bonded to the metallic raceway at one or both ends, as required.

quired. See Fig. 34. **Caution: Do not confuse the common grounding conductor that we are discussing here with the equipment grounding conductors.**

Due to corrosion, aluminum grounding conductors shall not be placed in direct contact with masonry, earth, or other corrosive materials. Also, where aluminum grounding conductors are used, they shall not be closer than 18 inches to the earth. Please note that this does not prohibit the use of aluminum grounding conductors, but the restrictions placed upon its use will practically eliminate its use, since it is not to be spliced.

Table 250-94 is used for sizing of the common grounding conductors in grounded systems. This will cover residential wiring.

K. Grounding Conductor Connections

To Raceway or Cable Armor. Care shall be taken to ground raceways or cable armor or interior wiring by connecting to grounding conductors as near as possible to the source of supply. Also, the grounding conductor shall be chosen so that no raceway or cable armor is grounded by a grounding conductor smaller than that required by **Table 250-95.** Refer to the NEC for this Table.

To Electrode. The grounding connection to the electrode shall be located as follows:

(a) **To Water Pipes.** System or common grounding conductors shall be attached to water pipes:

  (1) On the street side of the water meter.
  (2) To a cold water pipe of adequate current-carrying capacity.
  (3) As near as possible to the entrance of the water supply into the building.
  (4) If not on the street side of the water meter, the meter shall be adequately bonded and the bonding

jumper shall be long enough so that the water meter can be removed without disturbing the jumper. See Fig. 32.

(5) Care must be taken to place grounding conductors on the street or supply side of insulated couplings, unions, valves, etc.

(6) Attachment to water pipes shall be accessible if at all possible.

(7) If the water supply is from a well on the premises, the point of attachment shall be as close as possible to the well, with the casing bonded to the water piping. See Fig. 28.

(b) **To Gas Pipes.** Attachment to gas piping was covered in **Section 250-82**—recall that permission to make such an attachment is required. When connecting the ground to a gas pipe, it shall be on the street side of the meter and the connection shall be accessible if at all possible. Caution must be used as insulating couplings are often present in gas piping systems.

(c) **To Other Electrodes.** Grounding to electrodes other than metallic water piping was permitted in **Sections 250-82** and **83.** When grounding to this type of electrode, make certain that it is a permanent ground and that the connection is accessible if at all possible.

250-113. To Conductors and Equipment. The grounding conductor, bond, or bonding jumper shall be attached to circuits, conduits, cabinets, equipment, and the like, which are to be grounded, by means of:

(1) Suitable lugs.
(2) Pressure connectors.

(3) Clamps.

(4) Other approved means.

Soldering is never permitted. The strap type of grounding clamp is also not permitted. Be certain that the device used is approved for the purpose.

Table 250-94. Grounding Electrode Conductor for Grounded Systems

Size of Largest Service-Entrance Conductor or Equivalent for Parallel Conductors		Size of Grounding Electrode Conductor	
Copper	Aluminum or Copper-Clad Aluminum	Copper	*Aluminum or Copper-Clad Aluminum
2 or smaller	0 or smaller	8	6
1 or 0	2/0 or 3/0	6	4
2/0 or 3/0	4/0 or 250 MCM	4	2
Over 3/0 thru 350 MCM	Over 250 MCM thru 500 MCM	2	0
Over 350 MCM thru 600 MCM	Over 500 MCM thru 900 MCM	0	3/0
Over 600 MCM thru 1100 MCM	Over 900 MCM thru 1750 MCM	2/0	4/0
Over 1100 MCM	Over 1750 MCM	3/0	250 MCM

Where there are no service-entrance conductors, the grounding electrode conductor size shall be determined by the equivalent size of the largest service-entrance conductor required for the load to be served.

*See installation restrictions in Section 250-92(a).

See Section 250-23(b).

CHAPTER 6

Nonmetallic Sheathed Cable

(TYPES NM AND NMC)

Nonmetallic sheathed cable, as covered in **Article 336** or the NEC, is very widely used in the wiring of residences. It is probably the most widely used type of wiring for this purpose. NM cable is often referred to as ROMEX, loomwire, etc. These are nonmetallic sheathed cables approved in sizes No. 14 through No. 2 AWG with copper conductors and in sizes No. 12 through No. 2 AWG with aluminum conductors. By conductors is meant the current-carrying conductors in these cables. In addition to the two or three current-carrying conductors, the cable will also contain an equipment grounding conductor of proper size, as specified in **Table 250-95.** This grounding conductor may be bare or covered with green insulation, or green insulation with one or more yellow stripes. This grounding conductor is definitely not a current-carrying conductor. It is for grounding receptacle boxes, receptacles, switch boxes, and equipment of any nature, including enclosures, raceways, and appliances. See Fig. 35. This grounding conductor is an equipment grounding conductor, and the only time that it will carry current is when a fault develops in the circuit.

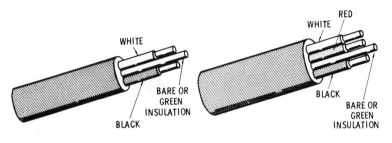

Fig. 35. NM, NMC, or UF cable.

Some years back, the grounding type receptacle, as shown in Fig. 36, appeared in the NEC. At this time there were no current using devices on the market which had cords that included an equipment grounding conductor and the grounding-type attachment plug as show in Fig. 37. Along

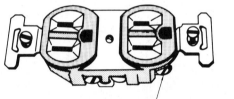

GREEN EQUIPMENT GROUNDING CONDUCTOR SCREW

Fig. 36. A grounding type receptacle.

with this came the NM Cable with ground for grounding purposes. Since the appearance of the grounding-type receptacles, there are a sufficient number of this type installed on grounded circuits that it became practical to make some changes in the 1968 Code—mainly to require the grounding of more appliances than were required to be grounded under previous Codes.

Prior to the appearance of this type of equipment grounding, it was necessary on certain appliances, such as an electric washing machine, to ground the appliance frame by means of a separate grounding conductor clamped to the cold-water piping. This was a rather ineffective grounding method because high impedances were often encountered.

Now with the grounding-type receptacles, NM cable with ground, and the spelling out of how the grounding is to be accomplished, we have a very effective equipment grounding circuit which, when properly installed, makes for a low impedance and thus for a more effectual means of operating the over-current devices or devices protecting the circuit in case of ground faults. A word of caution at this point—in wiring existing residences, you shall definitely not install a grounding-type receptacle on a circuit that does not have an equipment ground. This would create false security, and accidents would surely occur. The NEC has taken into account that we will run into existing ungrounded circuits on which extensions for additional receptacles must be made. This is

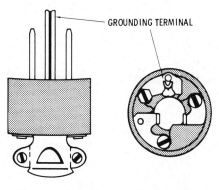

Fig. 37. A grounding type attachment plug.

covered in **Section 250-50(a)** and (**b**). Exception: For extensions only in existing installations which do not have a grounding conductor in the branch circuit, the grounding conductor of a grounding type receptacle outlet may be grounded to a grounded cold water pipe near the equipment. See Fig. 38. **Caution: This does not make the entire circuit a grounded circuit, but merely grounds the receptacles added to an existing circuit.**

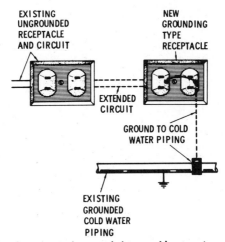

EXISTING
UNGROUNDED
RECEPTACLE
AND CIRCUIT

NEW
GROUNDING
TYPE
RECEPTACLE

EXTENDED
CIRCUIT

GROUND TO COLD
WATER PIPING

EXISTING
GROUNDED
COLD WATER
PIPING

Fig. 38. A grounding type receptacle grounded to a cold-water pipe when installed in an existing ungrounded system.

Sometimes in my duties as an inspector, I find that someone has been sold NM cable without ground. I will not permit an external green grounding conductor to be added, and my reason for this is that a sufficient amount of time has elapsed since the Code went to grounded circuits that this cable should not remain in existence. Also, for a grounding circuit to be effective, the impedance must be kept low, and I personally feel that to accomplish this the current-carrying conductors and the grounding conductor must be within the same common sheath.

Refer to **Section 250-45 (Equipment Connected by Cord and Plug.)** Portions of this Section will be quoted, but it would be well to read the entire Section in the NEC:

Under any of the following conditions, exposed non-current-carrying metal parts of cord and plug connected equipment, which are liable to become energized, shall be grounded:

(c) **In Residential Occupancies.** *In residential occupancies: (1) refrigerators, freezers, and air conditioners; (2) clothes-washing, clothes-drying, dish-washing machines, sump pumps, electrical aquarium equipment; (3) hand-held motor operated tools; (4) motor-operated appliances of the following types: hedge clippers, lawn mowers, snow blowers, and wet scrubbers; (5) portable hand lamps.*

Exception: Listed tools and listed appliances protected by a system of double insulation, or the equivalent, shall not be required to be grounded. Where such a system is employed, the equipment shall be distinctively marked.

Section 422-16 refers us to Article 250 for grounding of appliances and refers us to **Sections 250-42, 250-43,** and **250-45** for the grounding of refrigerators and freezers. Referral is made to **Sections 250-57** and **250-60** for grounding of clothes-washers, clothes-dryers, electric ranges, wall-mounted ovens, counter-mounted cooking units.

A portion of **Table 250-95**, which gives us the sizing for equipment grounding conductors, will be listed here. The entire table is not shown as the large ampacity circuits are not applicable to the conductors which we will find used in residences. Your attention should be called to the fact that the 1968 NEC made some changes in the sizes of equipment grounding conductors. Please note that for 15, 20 and 30 amp. circuits, the equipment grounding conductors are the same size as the phase conductors. Larger circuits shown permit a reduction in the size of the equipment grounding conductors.

At this point your attention is called to the fact that **Article 370** was amended to count the equipment ground-

Table 250-95. Size of Equipment Grounding Conductor for Grounding Interior Raceway and Equipment

Rating or Setting of Automatic Overcurrent Device in Circuit Ahead of Equipment, Conduit, etc., Not Exceeding (Amperes)	Size	
	Copper Wire No.	Aluminum Wire No.*
15	14	12
20	12	10
30	10	8
40	10	8
60	10	8
100	8	6
200	6	4
400	3	1
600	1	2/0
800	0	3/0
1000	2/0	4/0
1200	3/0	250 MCM
1600	4/0	350 "
2000	250 MCM	400 "
2500	350 "	500 "
3000	400 "	600 "
4000	500 "	800 "
5000	700 "	1000 "
6000	800 "	1200 "

* See installation restrictions in Section 250-92(a).

ing conductor in the fill of boxes. This will not be covered any further now, but will be covered fully a little later in this book.

USE AND INSTALLATION OF NM AND NMC CABLES

Section 336-3 gives us the information where nonmetallic sheathed cable may be used. Fundamentally, it may be used for both exposed and concealed installations.

For uses *not permissible* for either type NM or NMC cables, the following is listed:

(1) As service-entrance cable.

(2) In commercial garages.

(3) In theatres, except as provided in **Section 520-4** (this is a place of assembly).

(4) In motion picture studios.

(5) In storage battery rooms.

(6) In hoistways.

(7) In any hazardous location.

(8) Embedded in poured concrete, cement, or aggregate.

Type NM cable, as previously stated, may be installed for both exposed and concealed work in normally dry locations. This would include installing or fishing of NM cable in the air voids in masonry block or tile walls, but only if such walls are not exposed to excessive moisture or dampness. An example of type NM cable installation in masonry blocks would be in blocks used in the interior walls of buildings. Exterior walls of block (including exterior basement walls) shall not have NM cable installed in the voids of these blocks. See Fig. 39.

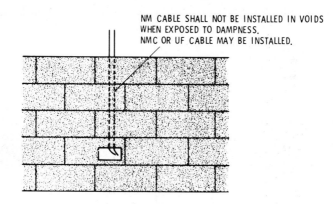

NM CABLE SHALL NOT BE INSTALLED IN VOIDS WHEN EXPOSED TO DAMPNESS. NMC OR UF CABLE MAY BE INSTALLED.

Fig. 39. NMC or UF cable may be installed in the voids of concrete-block walls, but not NM cable.

NMC cable is a moisture and corrosion-resistant type of cable. Therefore, NMC cables may be installed for both exposed and concealed work in any place that NM cable would be permitted, and may also be installed in moist, damp, or corrosive locations. Thus, it may be installed in the hollow voids of masonry block or tile walls used as outside walls of buildings, both above ground walls and basement walls. NMC cable may also be embedded in plaster or run in shallow chases in masonry walls and covered with plaster. When so installed, if this cable is within 2 inches of the finished surface, it shall be protected against damage from nails by covering the chase with a corrosion-resistant coated steel at least 1/16-inch thick and with a minimum width of ¾ inch. This protective metal covering may be under the final surface finish. See Fig. 40.

In the 1978 NEC there is an addition to restrictions on the use of NM cable. NM and NMC cables are not permitted to be used in more than two-family dwellings or in multifamily dwelling that exceed three floors above grade.

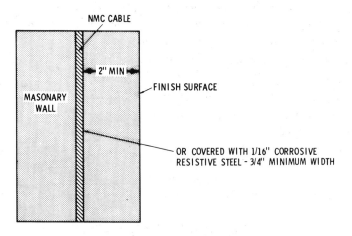

Fig. 40. Proper installation of NMC cable in a plastered wall.

NM and NMC cables shall be secured only with approved staples, straps, or similar fittings. These approved devices for the support of these cables are designed so as not to injure the cables.

These cables shall be secured within 12 inches of every cabinet, box, or fitting, and shall be secured at intervals of not over 4½ feet elsewhere. Of course, in wiring finished buildings, panels for prefabricated buildings, etc., the supports just mentioned are impractical, so we are permitted to fish these cables between access points. See Fig. 41.

It might be well to mention that should it be desired to use NM cable in the air voids in masonry block walls where,

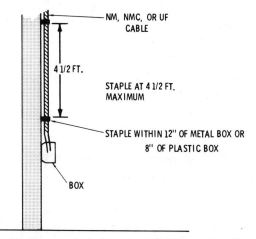

Fig. 41. Proper method of securing NM, NMC, or UF cable.

as previously stated, its use is prohibited, rigid conduit or EMT may be installed in the voids first and then NM cable fished through these conduits or EMT.

Section 336-6 is headed **Exposed Work-General; Section 336-8** is headed **In Unfinished Basements; Section 336-9** is headed **In Accessible Attics.** These three Sections will be covered at one time, as they are of a very similar nature.

In running NM or NMC cable in exposed work:

(a) **To Follow Surfaces.** The cable shall closely follow the surface of the building finish or of running boards.

(b) **Protection from Physical Damage.** It shall be protected from physical damage where necessary, by conduit, pipe, guard strips or other means. Where passing through a floor the cable should be enclosed in rigid metal conduit or metal pipe extending at least 6 inches above the floor.

This quote from the NEC is quite clear. However, it is illustrated in Fig. 43 to show what is meant in part (a), namely following the surface of the building. Fig. 44 illustrates part (b), where protection from physical damage is required. Fig. 45 illustrates where conduit, etc., is required when passing these cables through a floor.

Now refer to **Section 336-8 (In Unfinished Basements)**: *Where the cable is run at angles with joists in unfinished basements, assemblies not smaller than two No. 6 or three No. 8 conductors may be secured directly to the lower*

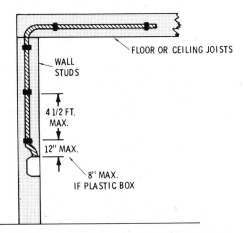

FLOOR OR CEILING JOISTS

WALL STUDS

4 1/2 FT. MAX.

12" MAX.

8" MAX. IF PLASTIC BOX

Fig. 43. Proper installation of exposed NM or NMC cable.

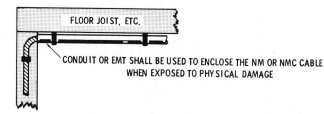

Fig. 44. Installation method to protect NM or NMC cable from physical damage.

edges of the joists: smaller assemblies shall either be run through bored holes in the joists or on running boards. Where run parallel to joists, cable size shall be secured to the sides or face of the joists. See Fig. 45, 46 and 47.

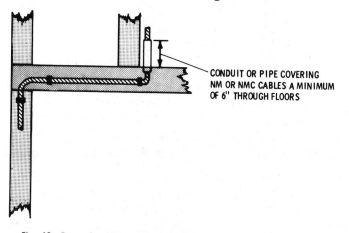

Fig. 45. Protecting NM or NMC cable where it passes through a floor.

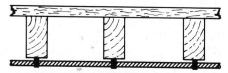

Fig. 46. Large NM or NMC cables may be run across the lower edges of joists in unfinished basements.

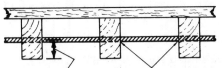

CABLES SMALLER THAN NO. 6-2 AND NO. 8-3 MAY
BE RUN THROUGH HOLES BORED THROUGH JOISTS

SHALL BE A MINIMUM OF
2 " or 1/16" STEEL PLATE
SHALL PROTECT CABLES

Fig. 47. Small NM or NMC cable must be run through bored holes in the joists in unfinished basements or protected by running boards.

Section 336-9. In Accessible Attics. *Cable in accessible attics or roof spaces shall also conform with* **Section 333-12.**

Going back to **Section 333-12. In Accessible Attics.** *Type AC cables in accessible attics or roof spaces shall be installed as follows:*

(a) *Where run across the top of floor joists, or within 7 feet of floor or floor joists across the face of rafters or studding, in attics and roof spaces which are accessible, the cable shall be protected by substantial guard strips which are at least as high as the cable. Where this space is not accessible by permanent stairs or ladders, protection will only be required within 6 feet of the nearest edge of scuttle hole or attic entrance.*

(b) *Where cable is carried along the sides of rafters, studs or floor joists, neither guard strips nor running boards shall be required.* See Figs. 48 and 49.

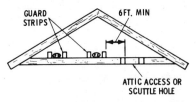

GUARD
STRIPS

6FT. MIN

ATTIC ACCESS OR
SCUTTLE HOLE

Fig. 48. Guard strips are used to protect cable in accessible attics under certain conditions.

90

Fig. 49. Cable run along the sides of rafters, studs, or joists in accessible attics need no additional protection.

Section 336-7. Through Studs, Joists and Rafters. See Section 300-4.

(a) Where exposed or concealed wiring conductors in insulating tubes or cables are installed through bored holes in studs, joists or similar wood members, holes shall be bored at the approximate centers of wood members, or at least 1¼ inches from the nearest edge where practical.

In analyzing (a), we find the 1¼-inch requirement from the nearest edge where practical. The intent of the statement *where practical* is in reference to a 2 x 4 stud, or similar building part, which in reality is only about 3⅝", but is considered to be 4" for all practical purposes. Bear in mind that, fundamentally, the 2" requirement is the prevailing factor.

(b) Where there is no objection because of weakening the building structure, metal-clad or nonmetallic-sheathed cable, aluminum-sheathed cable and Type MI cable may be laid in notches in the studding or joists when the cable at those points is protected against the driving of nails into it by having the notch covered with a steel plate at least ¹⁄₁₆ inch in thickness before building finish is applied. See Fig. 50.

Examining both (a) and (b) the primary purpose of all this is to protect the cables from being penetrated by nails or staples used in attaching lath, drywall, or panelling. Any one that has ever had to troubleshoot a case caused by

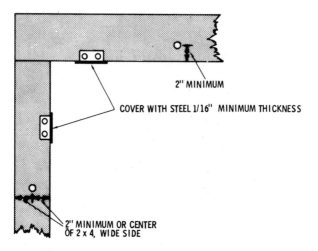

Fig. 50. Steel plates used to protect cable notched into studs or joists.

the penetration of a nail or staple into a cable will fully appreciate the intent of the NEC on this point.

In the installation of NM and NMC cables, **Section 336-10, Bends** shall be adhered to. This Section states: *Bends in cable shall be so made, and other handling shall be such, that the protective coverings of the cable will not be injured, and no bend shall have a radius less than 5 times the diameter of the cable.*

As an inspector, I have found damaged NM and NMC cables due to sharp bending, or damage to the other covering caused by pulling these cables through bored holes, etc. A recent case of trouble appeared in a two-section module home. This wiring was roughed-in before the inner and outer walls were attached, but on the final inspection, one branch circuit kept tripping the breaker. In running this fault down, it was found to be in the last portion of the branch circuit. The walls had to be removed and the cable was ordered to be replaced with a new length. When the cause of the fault was investigated, we found that at some time the cable had been bent very sharply, causing the insulation on one of the phase conductors to split and this phase conductor in turn was making contact with the bare equipment grounding conductor in the cable. While we were testing to find this fault, we did notice that by moving the cable, the fault would disappear and reappear. Too much emphasis cannot be placed on handling these cables properly during their installation. This type of fault may also occur by driving too severely the staples holding the cables.

CHAPTER 7

Underground Feeder and Branch Circuit Cable

(TYPE UF)

Artcle 339 of the NEC covers Type UF cable. This is an underground feeder or branch circuit cable of an approved type that comes in sizes No. 14 to 4/0 AWG, inclusive. The covering on the cable is:

1. Flame retardant
2. Moisture-resistant.
3. Fungus-resistant.
4. Corrosive-resistant.
5. Suitable for direct burial in the earth.

Section 339-3 is headed Use. We need not go into the entire section, but will only cover that portion that we might consider applicable to residential wiring.

Fundamentally, Type UF is a cable which may be buried directly in the earth, and may be made up of a single conductor or multiple conductors in one sheath. We will first discuss burying this cable in the earth.

Section 339-3. Use (3) refers us to Section 300-5. This refers us to Table 300-5, also the *Exceptions* under Section 300-5. Table under Section 300-5(a) gives us a minimum depth of 24 inches which shall be maintained for conductors and cables buried directly in the earth. This depth may be reduced to 12 inches provided supplemental protective covering such as a 2-inch concrete pad, metal raceway, pipe or other suitable protection is used. See Fig. 51.

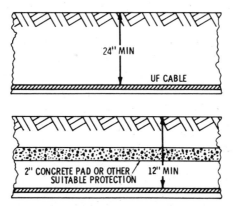

Fig. 51. Minimum depth at which UF cable must be buried.

Now let us analyze this portion of the NEC. We find the minimum depth of burial is 24 inches. It will be found that if this cable is buried in a rock type soil, the inspector will, without doubt, require the cable to have a fine sand bed placed in the bottom of the ditch with a layer of fine sand required as the first cover over this cable. This point is especially important in climates where the ground is subject to freezing to this depth or deeper, and the intent is to protect the cable insulation from damage due to the rocks in the soil. See Fig. 52.

This 24-inch depth may be reduced to 12 inches if there is a 2-inch concrete pad, metal raceway, etc., protecting the cable as called for. This cable must enter the ground at some point and must also emerge at some point. So, at these points, the UF cable must be sleeved in metal

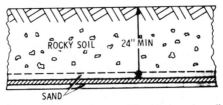

Fig. 52. Method of burying UF cable in rocky soil.

raceways and proper protection given so that frost heave will not cause cutting of the cable against the metal raceway. This can, of course, be done by using plastic insulating bushings at these points.

In **Section 339-3** we are told that this cable shall not be used when: (8) embedded in poured cement, concrete or aggregate, except where recognized in **Article 424.**

300-5(c). Underground Cables under Buildings. *Underground cable installed under a building shall be in raceway that is extended beyond the outside walls of the building.*

230-49. Protection against Damage—Underground. *Underground service conductors shall be protected against physical damage in accordance with* **Section 300-5.**

Let us stop a moment. Notice that this Section pertains to underground service conductors. However **Article 300,** headed WIRING METHODS-GENERAL, has **Section 300-5,** which is referred to in **Section 230-49.** Thus, even though **Section 230-49** is written to cover underground service conductors, it is also applicable to other conductors run underground. Now the following means are applicable to protecting UF cable:

(1) in duct;

(2) in rigid metal conduit or electrical metallic tubing made of a material suitable for the condition, or provided with corrosion protection suitable for the condition;

(3) in rigid nonmetallic conduit if installed in accordance with **Section 347-2** and **347-3;**

(4) in cable of one or more conductors approved for direct burial in the earth;

(5) other approved means.

We may ignore part (4) if we are talking about underground cable run under a building, as we are told in **Section 300-5** that it shall be in a raceway, etc.

In other words, any UF cable buried under a building must be installed so that it will be readily replaceable. You will also find that inspection authorities will, for the most part, consider concrete patios, etc., to be a part of the building. To clarify this, they would consider it impractical to put the portion of the UF cable under the building proper in a raceway and not have it in a raceway under a concrete patio.

There would be no way to replace the UF cable under the patio, should it become necessary to do so, without tunneling. So it makes good sense to require the raceway

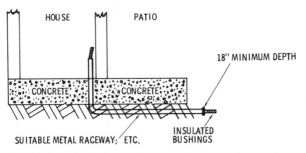

Fig. 53. UF cable buried under a building or other inaccessible locations must be installed so as to be readily replaceable.

out to the point where the cable may be reached by digging in exposed earth. See Fig. 53.

Part (d) of **Section 339-3** states: *Type UF cable may be used for interior wiring in wet, dry, or corrosive locations under the recognized wiring methods of this Code, and when installed as nonmetallic-sheathed cable it shall conform with the installation provisions of* **Article 336** *and shall be of the multiple conductor type. . . .*

For the balance of the coverage of Type UF cable for residential wiring it will be necessary to refer to Chapter 6 of this book. The installation of UF cable will be done in exactly the same way as the installation of NM or NMC cables. Refer back to Fig. 38 especially, where you will note that UF (as well as NMC cable) may be run in the voids of block walls exposed to dampness.

CHAPTER 8

Armored Cable

Article 333 covers ARMORED CABLE, Type AC cable. What we are interested in here is in regards to residential wiring with Type AC cable.

Definition. Type AC cable is a fabricated assembly of insulated conductors in a metallic enclosure. See **Section 333-4.**

Section 333-4. Construction. *Type AC cable shall be an approved cable with acceptable metal covering. The insulated conductors shall conform to* **Section 333-5.**

Type AC cables are branch-circuit and feeder cables with armor of flexible metal tape. Cables of the AC Type, except ACL, shall have an internal bonding strip of copper or aluminum, in intimate contact with the armor for its entire length. See Fig. 54.

In residential wiring we are principally interested in a would-be Type AC Series. Many of us knew this as BX. When BX was in existence, a ground fault in the cable often caused the exterior armor to heat, even to the extent of causing some fires. This is the same result as found in flexible metal conduit and was due to a high impedance developed by the spirally wound exterior armor. The NEC took this into account for flexible metal conduit. It required an equipment ground to be installed in the flexible metal conduit so that a low-impedance grounding circuit was obtained to take care of the heating and to provide a low enough impedance to actuate the overcurrent devices in the circuit. The NEC also made provisions that if approved fittings and flexible metal conduit were developed to take care of the impedance factor, the Code was prepared for its usage without a change. See Fig. 54.

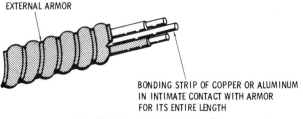

EXTERNAL ARMOR

BONDING STRIP OF COPPER OR ALUMINUM
IN INTIMATE CONTACT WITH ARMOR
FOR ITS ENTIRE LENGTH

Fig. 54. Type AC metal-clad cable.

In **Section 333-5,** *for cables of Type AC, insulated conductors shall be of a type listed in* **Table 310-13.** *In addition, the conductors shall have an over-all moisture-resistant and fire-retardant fibrous covering; . . .*

In **Section 333-6,** headed **Use,** find **Type AC.** Metal-clad cable of the AC type may be used in dry locations; for under plaster extensions as provided in **Article 333-6,** and embedded in plaster finish on brick or other masonry, except in damp or wet locations. This cable may be run or fished in the air voids of masonry block or tile walls; where such walls are exposed or subject to excessive moisture or dampness or are below grade line, Type ACL cable shall be used. This cable shall contain lead-covered conductors (Type ACL), if used where exposed to the weather or to continuous moisture, for underground runs and embedded in masonry, concrete or fill in buildings in course of construction or where exposed to oil, or other conditions having a deteriorating effect on the insulation. . . .

The difference between Types AC and ACL is that the conductors in Type ACL are lead covered. Type ACL is not used very often in residential wiring. However, Type AC may be used, but not where necessary to fish it through the air voids of masonry block or tile walls where these walls are below grade line or exposed or subject to excessive moisture or dampness. Refer to both Chapters 6 and 7 where NM cable was not permitted for that purpose, but NMC and Type UF were. Also, most inspection authorities would not look favorably on the installation of Type AC cable in chases of outside brick walls exposed to the dampness of weather conditions.

There need not be too much said about the installation of Type AC cable, as the requirements are very similar to those for Type NM covered in Chapter 6. It should be mentioned, however, that at each termination of AC cable, a fiber bushing made for this purpose shall be inserted between the armored sheath and the conductors. It should also be noted that the grounding strip, as referred to in the construction of Type AC cable, shall be either bent back over the fiber bushing and in intimate contact with the external portion of the armor and the cable clamp so installed that this grounding strip is made electrically secure by the clamp, or else this grounding strip shall be installed as was the equipment grounding conductor in NM cable. See Fig. 55.

The supports for Type AC cable are the same as for NM cable, namely at intervals not exceeding 4½ feet and within

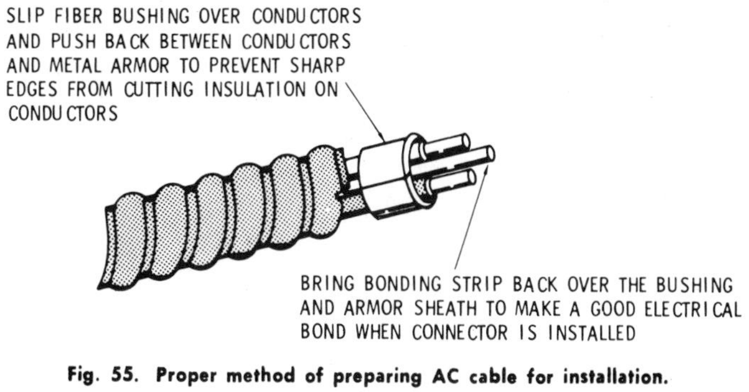

SLIP FIBER BUSHING OVER CONDUCTORS AND PUSH BACK BETWEEN CONDUCTORS AND METAL ARMOR TO PREVENT SHARP EDGES FROM CUTTING INSULATION ON CONDUCTORS

BRING BONDING STRIP BACK OVER THE BUSHING AND ARMOR SHEATH TO MAKE A GOOD ELECTRICAL BOND WHEN CONNECTOR IS INSTALLED

Fig. 55. Proper method of preparing AC cable for installation.

101

12 inches of every outlet box or fitting. Of course we cannot staple where Type AC cable is fished in walls. Also, lengths not to exceed 24 inches may be used at terminals if required. The boxes and fittings shall be approved for Type AC cable. All other conditions of installation shall be the same as for Types NM, NMC, and UF cables for the interior wiring of a residence, unless there are exceptions which were noted in the first part of this Chapter.

CHAPTER 9

Service-Entrance Cable

(TYPES SE AND USE)

Mention was made about service-entrance cable in Chapter 5. Not too much was covered, however, as it was felt that because SE and USE cables have various usages, it would be better to cover them in a separate chapter.

Section 338-1. Definition. Service-entrance cable is a conductor assembly provided with a suitable over-all covering, primarily used for services and of the following types. When consisting of two or more conductors, one may be without individual insulation. See Fig. 56.

(a) *Type SE, having a flame-retardant, moisture-resistant covering, but not required to have inherent protection against mechanical abuse.*

OUTER COVERING SHALL BE FLAME-RETARDANT
AND MOISTURE-RESISTANT ;
THERE SHALL BE MOISTURE SEAL TAPES UNDER
THIS OUTER COVERING.

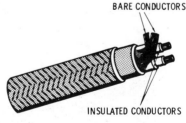

BARE CONDUCTORS

INSULATED CONDUCTORS

Fig. 56. SE cable with a bare neutral conductor.

(**b**) Type USE, recognized for underground use, having a moisture-resistant covering, but not required to have a flame-retardant covering or inherent protection against mechanical abuse. Single-conductor cables having rubber insulation specifically approved for the purpose do not require an outer covering.

Type SE cable is so designated because it is often used for service-entrance conductors. Refer back to Chapter 5 which covers in general details the installation of service entrances and service-entrance conductors.

Section 338-3, headed **Use as Branch Circuit or Feeders**, gives us more detailed information for use.

(a) **Grounded Conductor Insulated.** *Type SE, service-entrance cables may be used in interior wiring systems where all of the circuit conductors of the cable are of the rubber-covered or thermoplastic type.*

Take note of that portion which states *where all of the circuit conductors of the cable are of the rubber-covered or thermoplastic type.* Fig. 57 shows such a cable. The cable shown contains three insulated conductors to be used as circuit conductors, namely for wiring 115/230-volt systems which are the type usually used in residential wiring.

OUTER COVERING SHALL BE FLAME-RETARDANT AND MOISTURE-RESISTANT

THIS CABLE SHALL CONTAIN THREE INSULATED CONDUCTORS, ONE TO BE USED AS NEUTRAL, AND SHALL ALSO HAVE ONE STRANDED BARE CONDUCTOR TO BE USED AS AN EQUIPMENT GROUNDING CONDUCTOR

Fig. 57. SE cable with insulated conductors and an equipment grounding conductor.

Two of the insulated conductors are phase or current-carrying conductors, and the third insulated conductor is the neutral. There should also be one stranded bare conductor which is used as an equipment grounding conductor. In a majority of uses for branch circuits and feeders, the equipment grounding conductor will be required. Note that following part (b) of **Section 338-3** tells us: The above provisions do not intend to deny the use of service-entrance cable for interior use when the fully insulated conductors are used for circuit wiring and the uninsulated conductor is used for equipment grounding purposes.

An explanation of this statement is as follows, referring to the type SE cable illustrated in Fig. 56. For instance, a 230-volt branch circuit or feeder might be used to supply 230-volt loads which do not require a neutral conductor. These could be electrical heating panels, or electrical heating equipment. This would require the use of the two insulated conductors for the 230-volt supply and the bare conductor would be used as an equipment grounding conductor. You are cautioned to observe this information very closely, because for some reason the use of SE cables seems to be confusing at times.

Caution should be exercised when using the four-conductor type for a service entrance. In this type of cable there is an insulated neutral plus two phase conductors and a bare grounding conductor. The common practice seems to be to connect both the bare equipment grounding conductor and the insulated neutral conductor to the meter housing neutral terminal and at the neutral bus in the service equipment. This is a violation of **Section 310-4. (Conductors in Multiple)**. In this section is listed the criteria for paralleling conductors. This type of cable does not have a neutral conductor and an equipment grounding conductor which meets

105

the criteria of **Section 310-4.** Therefore, the bare equipment grounding conductor should be eliminated when this type of cable is used for service-entrance conductors.

Section 338-3(b) is now quoted. *Type SE, service-entrance cables without individual insulation on the grounded circuit conductor shall not be used as a branch circuit or as a feeder within a building, except a cable which has a final nonmetallic outer covering and when supplied by alternating current at not exceeding 150 volts to ground, may be used: (1) As a branch circuit to supply only a range, wall-mounted oven, counter-mounted cooking unit, or clothes dryer as covered in* **Section 250-60,** *or (2) as a feeder to supply only other buildings on the same premises.* It shall not be used as a feeder terminating within the same building in which it originates.

We find some of the same material here that was previously explained in this chapter. In (1), we find that a cable with insulated conductors and one bare conductor, as illustrated in Fig. 56, may be used as a branch circuit to supply a range, wall-mounted oven, counter-mounted cooking unit, or clothes dryer. Reference is made to **Section 250-60** which tells us that a three conductor cable (SE), with one con-

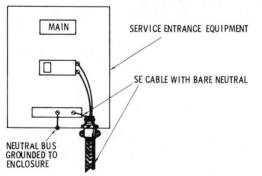

Fig. 58. SE cable with a bare neutral conductor may be used as a branch circuit to supply ranges, dryers, etc., only when it originates in service equipment panels.

ductor bare, may be used to supply ranges, etc., only when this branch circuit originates in service equipment panels. See Fig. 58. In case this is confusing, the reasoning is as follows:

In the service equipment, the neutral bus is to be grounded to the panel enclosure. Therefore, in using the bare neutral of the Type SE cable under discussion, hazardous conditions are created, because should the bare neutral happen to touch the enclosure, they are electrically connected together in the service equipment. When supplying ranges, etc., from a feeder panel, however, an entirely different condition exists. The neutral bus in the feeder panel is isolated from the enclosure, as shown in Figs. 59 and 60. In Fig. 59 the service equipment and a feeder panel is supplied by a four-conductor SE cable having three insulated conductors and one

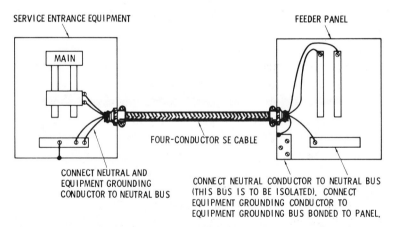

SERVICE ENTRANCE EQUIPMENT   FEEDER PANEL

MAIN

FOUR-CONDUCTOR SE CABLE

CONNECT NEUTRAL AND
EQUIPMENT GROUNDING
CONDUCTOR TO NEUTRAL BUS

CONNECT NEUTRAL CONDUCTOR TO NEUTRAL BUS
(THIS BUS IS TO BE ISOLATED). CONNECT
EQUIPMENT GROUNDING CONDUCTOR TO
EQUIPMENT GROUNDING BUS BONDED TO PANEL.

Fig. 59. SE cable with three insulated conductors and a bare equipment grounding conductor must be used when supplying a feeder panel.

bare equipment grounding conductor. The neutral bus is bonded to the service-equipment enclosure and, in addition, the insulated neutral and the bare equipment grounding conductor of the SE cable are both tied to the neutral bus. The

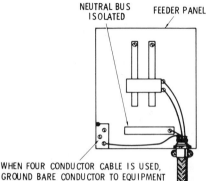

NEUTRAL BUS
ISOLATED

FEEDER PANEL

WHEN FOUR CONDUCTOR CABLE IS USED,
GROUND BARE CONDUCTOR TO EQUIPMENT
GROUNDING BUS. THIS SHALL BE BONDED
TO ENCLOSURE.

FOUR-CONDUCTOR CABLE OR THREE-CONDUCTOR
CABLE WITH ALL THREE CONDUCTORS INSULATED
TO SUPPLY RANGE OR DRYER FROM A FEEDER PANEL

Fig. 60. Branch circuit connections from a feeder panel.

neutral bus is isolated in the feeder panel and the insulated
neutral goes to this neutral bus. Notice that the grounding
bus in the service-entrance enclosure is tied to the enclosure,
and it is to this bus that the bare equipment grounding conductor is connected. In case someone thinks this is parallel-
ling the neutral and the equipment grounding conductor, it
should be stated that this is not true. Granted that these two
conductors are tied together in the service equipment enclosure, but they are not tied together in the feeder panel and
are therefore not paralleled.

In Fig. 60, either a four-conductor cable or a three-conductor cable with all three conductors insulated is used as the
branch circuit to a range or dryer. What do we have? With
a four-conductor cable the neutral and the equipment ground
conductor are separated, but the neutral is an insulated conductor so that if it comes in contact with the enclosure
(which it most likely will), the purpose of isolating the
neutral bus in a feeder panel is not defeated. If a three-conductor insulated SE cable is used, the Code permits us to

ground the frame of the range or dryer to the neutral in residences. **Caution: If a four-conductor cable (one conductor being bare) is installed, use the bare conductor for grounding the frame of the range or dryer, but do not ground the neutral to the frame also.** To do so would defeat the purpose we are after.

Part (2) of **Section 338-3(b)**, states that SE cable with a bare neutral, as illustrated in Fig. 56, may be used *as a feeder to supply only other buildings on the same premises.* Note specifically that it goes on to say that: *it shall not be used as a feeder terminating within the same building in which it originates.* This part stems from **Section 250-24**, headed **Two or More Buildings Supplied From Single-Service Equipment.** In a residence, we might have a detached garage. This could be supplied from the panel in the house by means of a feeder circuit using SE cable with two insulated conductors and a bar neutral but with no equipment grounding conductor. This is confusing because, in dealing with feeders, we normally think of an equipment grounding conductor having to be supplied with the feeder. However, in **Section 250-24** we find that we have a different situation. Where more than one building is supplied by the same service, the grounded circuit conductor of the wiring system of any building utilizing one branch circuit supplied from such service may be connected to a grounding electrode at such building, and in the case of any building housing equipment required to be grounded or utilizing two or more branch circuits supplied from such service, and in the case of a building housing live stock, shall be so connected. Thus, even though we have a feeder circuit, the other building is treated as a service for wiring purposes and a ground electrode is provided at the second building.

A word of caution is given at this point on checking the markings on SE cable. The outer surface of SE cable shall be marked at intervals not exceeding 24 inches. This marking shall designate the type of cable, the size of the conductors, and the type of insulation which is used on these conductors. This is important when using **Table 310-16** to determine the allowable ampacities of conductors. In this table we find that No. 8 aluminum conductors with 60°C insulation have an ampacity of 30, while those with 75°C insulation have an ampacity of 40, etc., so the overcurrent protection will have to fit this table.

When installing SE and USE cable, USE cable shall be treated in the same manner as UF cables. SE cable comes under the provisions of not only **Article 338** but also **Article 300** and **Article 336**. Most of these requirements were covered in Chapters 6, 7, and 8, so they will not be repeated at this point.

CHAPTER 10

Electrical Metallic Tubing

Electrical metallic tubing, commonly known as EMT or thinwall conduit, appears in **Article 348** of the NEC. Next to nonmetallic sheathed cable, which was covered in Chapter 6, the next most widely used wiring method for residential wiring is probably EMT. In discussing wiring with EMT, we will find that many of the requirements also include rigid metallic conduit.

EMT is now quite generally used most any place in residential wiring. There are a few places, however, where it is prohibited. EMT may be used where it will not be subject to severe physical damage both during and after installation. This takes care of most phases of residential wiring. However, please direct your attention to the second paragraph of **Section 348-1. Use.**

Unless made of a material judged suitable for the condition, or unless corrosion protection approved for the condition is provided, ferrous or nonferrous electrical metallic tubing, elbows, couplings and fittings shall not be installed in concrete or in direct contact with the earth, or in areas subject to severe corrosive influences. This automatically places the judging of conditions where EMT might or might not be installed in the earth or in concrete up to the authority having jurisdiction. You will find that most inspection authorities will never permit EMT in contact with the earth,

111

and under **Section 90-4** it becomes their right to judge the conditions. You will also find some inspection authorities who never permit EMT in concrete slabs, while others will permit such an installation provided that the concrete does not have an additive such as calcium chloride.

EMT is manufactured in sizes from ½″ through 4″ electrical trade size. Tubing smaller than ½″ shall only be used for underplaster extensions as covered in **Article 344** or for enclosing the leads of motors as permitted in **Section 430-145(b)**. In my years of experience, I have never come across EMT smaller than ½″ and I presume that this is because, from an economic standpoint, it is not practical to make smaller sizes commercially.

The number of conductors allowed in EMT is covered in **Section 348-6.** This section refers back to **Section 346-6.**

346-6. Number of Conductors in Conduit. The number of conductors permitted in a single conduit shall be as follows:

Where the conductors are all of the same size use **Tables 3A, 3B** and **3C** of Chapter 9 of the NEC. Where conductors are of varying sizes use **Tables 1, 4** and **5** of Chapter 9 of the NEC.

Tables 3A, 3B and **3C** in Chapter 9 of the NEC is reproduced here. Only that part that is applicable to residential wiring need be used.

Notice that this table includes a column A and a column B. With the development of the new type of insulation permitting a smaller overall size for certain conductors, it was felt that a distinction should be given in this table between the larger diameter and smaller diameter conductors.

Your attention is called to the note at the head of this table: *Derating factors for more than three conductors in raceways, see Note 8,* **Table 310-16** *through* **310-19.**

Referring to Note 8, we find:

8. More Than Three Conductors in a Raceway or Cable.
Tables 310-16 and **310-18** give the allowable ampacities for
not more than three conductors in a raceway or cable.
Where the number of conductors in a raceway or cable exceeds three, the allowable ampacity of each conductor shall
be reduced as shown in the following table:

Number of Conductors	Per Cent of Values in Tables 310-12 and 310-14
4 to 6	80
7 to 24	70
25 to 42	60
43 and above	50

Exception. No. 1—When conductors of different systems,
as provided in **Section 300-3,** *are installed in a common race-*
way the derating factors shown above apply to the number
of Power and Lighting (**Articles 210, 215, 220** *and* **230** *) con-*
ductors only.

To explain Note 8, turn to **Table 310-16** and use the 75°C
column for the proper conductors. Using No. 14 AWG, we
find that it has an allowable ampacity of 15 amperes. According to Note 8, we find that if we do not have over three
No. 14s in a raceway, we may use 15-ampere overcurrent
devices on this branch circuit, but if we have four No. 14s
in a raceway, we have to derate to 80% of the 15 amperes,
which would be 12 amperes. Since this is an impractical
combination, we change to No. 12 AWG which, according
to **Table 310-16,** gives us an allowable ampacity of 20

Table 3A. Maximum Number of Conductors in Trade Sizes of Conduit or Tubing (Based on Table 1, Chapter 9)

Type Letters	Conductor Size AWG, MCM	½	¾	1	1¼	1½	2	2½	3	3½	4	4½	5	6
TW, T, RUH, RUW, XHHW (14 thru 8)	14	9	15	25	44	60	99	142	171	176	108			
	12	7	12	19	35	47	78	111	131	84				
	10	5	9	15	26	36	60	85	62					
	8	2	4	7	12	17	28	40						
RHW and RHH (without outer covering), THW	14	6	10	16	29	40	65	93	143	192	163	106	133	
	12	4	8	13	24	32	53	76	117	157	85			
	10	4	6	11	19	26	43	61	95	127				
	8	1	3	5	10	13	22	32	49	66				
TW, T, THW, RUH (6 thru 2), RUW (6 thru 2)	6	1	2	4	7	10	16	23	36	48	62	78	97	141
	4	1	1	3	5	7	12	17	27	36	47	58	73	106
	3	1	1	2	4	6	10	15	23	31	40	50	63	91
	2	1	1	2	4	5	9	13	20	27	34	43	54	78
	1		1	1	3	4	6	9	14	19	25	31	39	57
FEPB (6 thru 2), RHW and RHH (without outer covering)	0		1	1	2	3	5	8	12	16	21	27	33	49
	00		1	1	1	3	5	7	10	14	18	23	29	41
	000		1	1	1	2	4	6	9	12	15	19	24	35
	0000			1	1	1	3	5	7	10	13	16	20	29

Table 3A. Maximum Number of Conductors in Trade Sizes of Conduit or Tubing (Based on Table 1, Chapter 9)

Type Letters	Conductor Size AWG, MCM	1/2	3/4	1	1 1/4	1 1/2	2	2 1/2	3	3 1/2	4	4 1/2	5	6
	250			1	1	1	2	4	6	8	10	13	16	23
	300			1	1	1	2	3	5	7	9	11	14	20
	350				1	1	1	3	4	6	8	10	12	18
	400				1	1	1	2	4	5	7	9	11	16
	500				1	1	1	1	3	4	6	7	9	14
	600					1	1	1	3	4	5	6	7	11
	700					1	1	1	2	3	4	5	7	10
	750					1	1	1	2	3	4	5	6	9

Table 3B. Maximum Number of Conductors in Trade Sizes of Conduit or Tubing (Based on Table 1, Chapter 9)

Type Letters	Conductor Size AWG, MCM	1/2	3/4	1	1 1/4	1 1/2	2	2 1/2	3	3 1/2	4	4 1/2	5	6
THWN,	14	13	24	39	69	94	154							
	12	10	18	29	51	70	114	164						
	10	6	11	18	32	44	73	104	160					
	8	3	5	9	16	22	36	51	79	106	136			
THHN,	6	1	4	6	11	15	26	37	57	76	98	125	154	
	4	1	2	4	7	9	16	22	35	47	60	75	94	137
FEP (14 thru 2),	3	1	1	3	6	8	13	19	29	39	51	64	80	116
FEPB (14 thru 8),	2	1	1	3	5	7	11	16	25	33	43	54	67	97
	1		1	1	3	5	8	12	18	25	32	40	50	72
XHHW (4 thru 500MCM)	0		1	1	3	4	7	10	15	21	27	33	42	61
	00		1	1	2	3	6	8	13	17	22	28	35	51
	000		1	1	1	3	5	7	11	14	18	23	29	42
	0000			1	1	2	4	6	9	12	15	19	24	35
	250			1	1	1	3	4	7	10	12	16	20	28
	300			1	1	1	3	4	6	8	11	13	17	24
	350				1	1	2	3	5	7	9	12	15	21
	400				1	1	1	3	5	6	8	10	13	19

Table 3B. Maximum Number of Conductors in Trade Sizes of Conduit or Tubing (Based on Table 1, Chapter 9)

Type Letters	Conductor Size AWG, MCM	1/2	3/4	1	1 1/4	1 1/2	2	2 1/2	3	3 1/2	4	4 1/2	5	6
	500				1	1	1	2	4	5	7	9	11	16
	600				1	1	1	1	3	4	5	7	9	13
	700					1	1	1	3	4	5	6	8	11
	750					1	1	1	2	3	4	6	7	11
XHHW	6	1	3	5	9	13	21	30	47	63	81	102	128	185
	600				1	1	1	1	3	4	5	7	9	13
	700					1	1	1	3	4	5	6	7	11
	750					1	1	1	2	3	4	6	7	10

Table 3C. Maximum Number of Conductors in Trade Sizes of Conduit or Tubing
(Based on Table 1, Chapter 9)

Type Letters	Conductor Size AWG, MCM	1/2	3/4	1	1 1/4	1 1/2	2	2 1/2	3	3 1/2	4	4 1/2	5	6
RHW,	14	3	6	10	18	25	41	58	90	121	155			
	12	3	5	9	15	21	35	50	77	103	132			
	10	2	4	7	13	18	29	41	64	86	110	138		
	8	1	2	4	7	9	16	22	35	47	60	75	94	137
RHH	6	1	1	2	5	6	11	15	24	32	41	51	64	93
	4	1	1	1	3	5	8	12	18	24	31	39	50	72
(with	3	1	1	1	3	4	7	10	16	22	28	35	44	63
outer	2		1	1	3	4	6	9	14	19	24	31	38	56
covering)	1		1	1	1	3	5	7	11	14	18	23	29	42
	0		1	1	1	2	4	6	9	12	16	20	25	37
	00			1	1	1	3	5	8	11	14	18	22	32
	000			1	1	1	3	4	7	9	12	15	19	28
	0000				1	1	2	4	6	8	10	13	16	24

Table 3C. Maximum Number of Conductors in Trade Sizes of Conduit or Tubing (Based on Table 1, Chapter 9)

Conduit Trade Size (Inches) Conductor Size AWG, MCM	1/2	3/4	1	1¼	1½	2	2½	3	3½	4	4½	5	6
250				1	1	1	3	5	6	8	11	13	19
300				1	1	1	3	4	5	7	9	11	17
350				1	1	1	2	4	5	6	8	10	15
400				1	1	1	1	3	4	6	7	9	14
500				1	1	1	1	3	4	5	6	8	11
600					1	1	1	2	3	4	5	6	9
700					1	1	1	1	3	3	4	6	8
750						1	1	1	3	3	4	5	8

Type Letters

Table 1. Percent of Cross Section of Conduit and Tubing for Conductors
(See Table 2 for Fixture Wires)

Number of Conductors	1	2	3	4	Over 4
All conductor types except lead-covered (new or rewiring)	53	31	40	40	40
Lead-covered conductors	55	30	40	38	35

Note 1. See Tables 3A, 3B and 3C for number of conductors all of the same size in trade sizes of conduit ½ inch through 6 inch.

Note 2. For conductors larger than 750 MCM or for combinations of conductors of different sizes, use Tables 4 through 8, Chapter 9, for dimensions of conductors, conduit and tubing.

Note 3. Where the calculated number of conductors, all of the same size, include a decimal fraction, the next higher whole number shall be used where this decimal is 0.8 or larger.

Note 4. When bare conductors are permitted by other Sections of this Code, the dimensions for bare conductors in Table 8 of Chapter 9 shall be permitted.

Note 5. A multi-conductor cable of three or more conductors shall be treated as a single conductor cable for calculating percentage conduit fill area.

amperes. Since we are using 4 conductors in a raceway, Note 8 tells us we must derate to 80% of the 20 amperes, which equals 16 amperes. Therefore, we will be required to use No. 12 copper wire on this 15-ampere branch circuit. The same type of calculations would be used for aluminum conductors except **Table 310-18.** would be used instead of **Table 310-16.**

Your attention is called to the Note at the head of both **Tables 310-16** and **310-18:** *Not More than Three Conductors in Raceway or Cable or Direct Burial (Based on Ambient Temperature of 30°C. 86°F.)*

Ambient temperature refers to the temperature of the surrounding area in which the conductors and raceways are installed. **Tables 310-16** through **310-19,** each have derating tables at the bottom of the ampacity Tables.

Section 348-7 tells us that EMT is not to be coupled together nor connected to boxes or fitting by means of threads made in the wall of the EMT. Nothing but approved fittings for the purpose are to be used. In using EMT connectors and/or couplings, be certain that they have the UL letters on them or at least on the boxes in which they come.

Fig. 61. An indent-type EMT connector.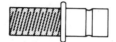

Fig. 61 shows an indent-type connector—there are also indent type couplings. When installing an indent-type of fitting, use the tool made for these fittings, and be very sure that it is not sprung or worn so that the indents do not make proper electrical continuity between the fitting and the EMT.

Fig. 62 shows a pressure-type coupling. Each end has a nut that, when screwed down on the coupling threads, puts pressure against a spring device within the nut so as to make good electrical contact.

Fig. 63 shows a setscrew-type connector. Couplings of this type are also available. Note the setscrew which is tightened down against the EMT inserted into the coupling or connector, thus making a good electrical contact.

EMT couplings and connectors shall be made tight for good electrical continuity. When buried in masonry or concrete, they shall be of the concrete-tight type, and when used in wet locations they shall be of the raintight type. A description of the purposes and locations where they may be used will be found on the cartons in which they are packed.

121

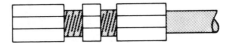

Fig. 62. A pressure-type EMT coupling.

Section 348-10. Bends—Number in One Run. A run of electrical metallic tubing between outlet and outlet, between fitting and fitting, or between outlet and fitting, shall not contain more than the equivalent of four quarter bends

Fig. 63. A setscrew-type EMT connector.

(360 degrees, total), including those bends located immediately at the outlet or fitting.

In Fig. 64 will be seen two boxes connected by a raceway with four 90-degree bends. This makes a total of 360° which is the maximum permitted by the Code. Fig. 65 also shows a total of 360° total bends between the boxes. This, however, differs from Fig. 64 in that two 45° offset bends were made at each box to bring the EMT from the knockout in the box to where it could be strapped to a wall. We also find two 90° bends, so we have used the allowable total number of bends permitted between boxes. The purpose of this 360° total is that it was apparently felt that more bends than this total amount would cause difficulty in pulling conductors into the raceway and possibly cause damage to the insulation.

Section 348-9. Bends—How Made. *Bends in the tubing shall be so made that the tubing will not be injured and that the internal diameter of the tubing will not be effect-*

ively reduced. The radius of the curve of the inner edge of any field bend shall not be less than shown in **Table 346-10**.

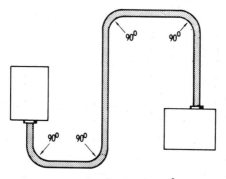

$$90^0 + 90^0 + 90^0 + 90^0 = 360^0$$

Fig. 64. The maximum amount of bends in EMT must not exceed 360° between any two boxes or pull points.

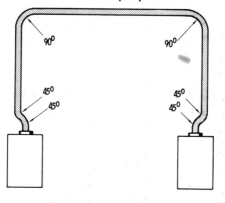

$$45^0 + 45^0 + 90^0 + 90^0 + 45^0 + 45^0 = 360^0$$

Fig. 65. Another example of the maximum amount of bends permitted in EMT between junctions.

Exception: For field bends made with a bending machine designed for the purpose, the minimum radius shall not be less than indicated in **Table 346-10** *Exception.*

We are shown how to measure the radius of a bend in Fig. 66.

There are many manufacturers that build benders of various types for EMT. Some are hand-type benders and some are hydraulic. Care must be taken to select the proper size bender to match the EMT being bent. Do not substitute a bender for rigid metal conduit when bending EMT. One case is recalled where 2″ EMT required two 45° offsets. The contractor did not have the proper bending equipment, nor did he take the time to borrow the proper equipment. Instead, he took the EMT to a blacksmith shop, and had it heated to make the bends. In doing this, the outer and inner coatings of the EMT were melted away causing roughness inside and corrosion on the outside. Of course this piece of EMT would not pass inspection and he was required to install a new piece of EMT bent by approved methods. The walls of EMT are thin and are very easily wrinkled or kinked, thus reducing the inside diameter. See Fig. 67.

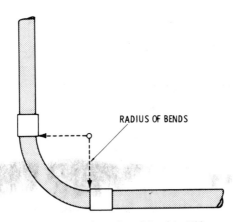

RADIUS OF BENDS

Fig. 66. The radius of bend in EMT.

Table 346-10. Radius of Conduit Bends (Inches)

Size of Conduit (In.)	Conductors Without Lead Sheath (In.)	Conductors With Lead Sheath (In.)
½	4	6
¾	5	8
1	6	11
1¼	8	14
1½	10	16
2	12	21
2½	15	25
3	18	31
3½	21	36
4	24	40
4½	27	45
5	30	50
6	36	61

Table 346-10. Exception. Radius of Conduit Bends (Inches)

Size of Conduit (In.)	Radius to Center of Conduit (In.)
½	4
¾	4½
1	5¾
1¼	7¼
1½	8¼
2	9½
2½	10½
3	13
3½	15
4	16
4½	20
5	24
6	30

Section 348-11. Reaming. *All cut ends of electrical metallic tubing shall be reamed to remove rough edges.*

125

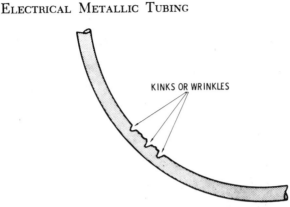

Fig. 67. Improper bending may produce wrinkles or kinks in EMT.

In Fig. 68, note that when EMT is cut with a wheel-type pipe cutter, the tubing tends to roll in. All of this must be cut away by means of a reamer. EMT cut with a hacksaw does not have this rolled-in burr, but care must be taken to make square cuts and then ream to eliminate any burrs and to give a beveled edge. It is impossible for an inspector to see every cut on a wiring job, but be assured that if this

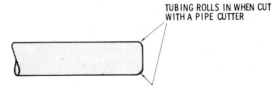

Fig. 68. Cutting EMT with a pipe cutter produces a rolled-in burr.

reaming is not done properly, breakdown or damage to the insulation will occur.

Section 348-12. Supports. *Electrical metallic tubing shall be installed as a complete system as provided in* **Article 300** *and shall be securely fastened in place at least every 10 feet and within 3 feet of each outlet box, junction box, cabinet, or fitting.* See Fig. 69.

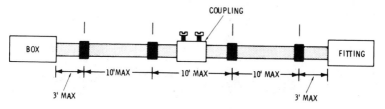

Fig. 69. EMT must be supported at specified maximum intervals.

Section 348-13 covers boxes and fittings which will be discussed in a later Chapter.

CHAPTER 11

Rigid Metal Conduit

Rigid metal conduit is covered in **Article 346** of the NEC. Much of the preceding chapter and **Article 348** of the NEC is also quite applicable to rigid metal conduit.

Rigid metal conduit may be used in all atmospheric conditions and occupancies except that ferrous metal raceways and fittings protected only by enamel may be used only indoors and in occupancies not subject to a corrosive influence. We must also bear in mind galvanic action (electrolysis) and its effect on dissimilar metals.

The paragraph (c) of **Section 346-1** is practically identical to **Section 348-1.** This paragraph, however, is very important and will be referred to here: Unless made of a material judged suitable for the condition, or unless corrosion protection approved for the condition is provided, ferrous or nonferrous metallic conduit, elbows, couplings and fittings shall not be installed in concrete or in direct contact with the earth, or in areas subject to severe corrosive influences.

This paragraph automatically places the judging of conditions and locations where rigid metal conduit may or may not be installed in earth or in concrete up to the authority having jurisdiction.

Conduit is available that is factory coated. There are also materials available which may be sprayed or painted on the conduit and which will be satisfactory for approval. Check with the inspection authority and see what they will require.

You will find that, to a large degree, inspection authorities will never permit rigid metal conduit in contact with the earth, and under **Section 90-4** it becomes their right to judge the conditions. However, you will find that most inspection authorities will permit rigid metal conduit in concrete without additional protection from corrosion unless the concrete has an additive such as calcium chloride. An inspector recently observed rigid galvanized conduit which had been in a concrete slab for several years. Calcium chloride had originally been added to the concrete to prevent freezing. When this concrete was broken up, it was found that the conduit was very badly eaten away. Practically no inspector will permit aluminum conduit in the earth or in concrete, as it seems to be especially vulnerable to corrosion in these areas. In deference to aluminum conduit, however, there are corrosive influences (but not in residential wiring locations) that do not affect aluminum nearly as much as they affect rigid galvanized conduit.

Section 346-5. Minimum Size. No conduit smaller than ½ inch, electrical trade size, shall be used, except as provided for underplaster extensions in **Article 344,** and for enclosing the leads on motors as permitted in **Section 430-145(b).**

Section 346-6. Number of Conductors in Conduit. Refer to the coverage given in Chapter 10 covering EMT. Everything stated there, including references to the tables, are all applicable to rigid metal conduit.

Section 346-7. Reaming. Same as reaming of EMT.

Section 346-8. Bushings. *Where a conduit enters a box or other fitting, a bushing shall be provided to protect the wire from abrasion unless the design of the box or fitting is such as to afford equivalent protection.* See **Section 373-6(c)** *for the protection of conductors at bushings.*

129

Section 373-6. Deflection of Conductors.

(c) **Insulated Bushings.** *Where ungrounded conduct-ors of No. 4 or larger enter a raceway in a cabinet, pull box, junction box, or auxiliary gutter, the conductors shall be protected by a substantial bushing providing a smoothly rounded insulating surface, unless the conductors are sep-arated from the raceway fitting by substantial insulating material securely fastened in place. Where conduit bushings are constructed wholly of insulating material, a locknut shall be installed both inside and outside the enclosure to which the conduit is attached.*

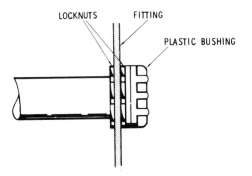

LOCKNUTS   FITTING

PLASTIC BUSHING

Fig. 70. Rigid metal conduit terminated at a junction box by locknuts and an insulated bushing.

Fig. 70 shows insulated bushings and locknuts to be used for this purpose. Fig. 71 illustrates a fiber insert to be used with standard metal bushings. Where bonding-type bush-ings are to be used, they can be metal with an insert, or they may be metal with a plastic throat.

Section 346-9. Couplings and Connectors.

(a) *Threadless couplings and connectors used with con-duit shall be made tight. Where buried in masonry or con-*

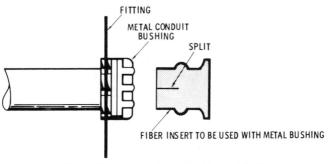

Fig. 71. A fiber insert to be used with metal bushings.

crete, they shall be of the concrete-tight type, or where installed in wet locations, shall be of the raintight type.

(**b**) *Running threads shall not be used on conduit for connection at couplings.*

For those who are not familiar with the term, *running threads,* these were previously of a form that might be called a union, using water-pipe terminology. This was accomplished by threading the conduit far enough back so that a locknut could be screwed on in addition to conduit coupling, which does not have tapered threading. Another

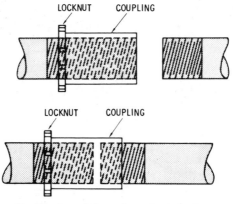

Fig. 72. A conduit coupling using locknuts.

131

piece of conduit was then screwed into the coupling, butting up against the first piece of conduit. This assembly was then locked by tightening the locknut which had previously been put on to the running thread. See Fig. 72.

Fig. 73 illustrates a fitting which serves the same purpose on conduit that a union does on water pipe. It is often called an Erickson.

Pipe unions are not approved for conduit use. In fact, neither water pipe or fittings is approved in place of rigid conduit. The interior coating of water pipe might be rough and damage the wire insulation. Attempts are often made to use water pipe and fittings for electrical raceways, even water pipe sections welded together. When found, they have to be turned down and approved raceways and fittings used to replace them.

Section 346-10. Bends—How Made. This was extensively covered in Chapter 10. Everything covered there is applicable to rigid metal conduit.

Section 346-12. Supports. *Rigid metal conduit shall be installed as a complete system as provided in* **Article 300** *and*

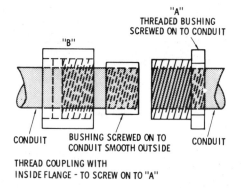

Fig. 73. A type of conduit coupling that serves as a union.

shall be securely fastened in place. Conduit shall be firmly fastened within 3 feet of each outlet box, junction box, cabinet, or fitting. Conduit shall be supported at least every ten feet except that straight runs of rigid conduit made up with approved threaded couplings may be secured in accordance with **Table 346-12,** *provided such fastening prevents transmission of stresses to terminus when conduit is deflected between supports.*

Your attention is called to that portion of **Section 346-12** which states: *conduit shall be supported at least every ten feet except that straight runs of rigid conduit made up with approved threaded couplings, etc.* Notice that it states *approved threaded couplings.* This definitely does not mention threadless couplings and connectors discussed in **Section 346-9.**

Last, but by no means least, all connections into threaded hubs, locknuts, and bushings, shall be made up tight. The rigid conduit is accepted as the equipment grounding means, and if these are not made up tight, electrical continuity will not be good. Thus, in case of a ground fault, high impedance will result and thus impede the operation of the overcurrent devices. There may also be arcing at the connection to boxes, cabinets, etc., and there might be a resulting fire. A good conduit job is a beauty to behold—a poor job is something else.

CHAPTER 12

Circuits Required for Dwellings

In Chapter 3, actual calculations for dwellings were given. Referring to Calculation No. 1, we found that we had 21,531 watts which we divided on a 3-wire single-phase basis into the total amperes per phase of 96 amperes. Further in this calculation it was determined that we had a general lighting load of 8874 watts which would be on 115-volt circuits. The 8874 watts was divided by 115 volts, which gave us 77.2 amperes for general lighting to be handled by a minimum of six 15-ampere circuits or four 20-ampere circuits. This included lighting and general-purpose receptacles and is the minimum number of circuits required. As stated before, do not figure the bare minimum, but add a few additional circuits.

To the general lighting and receptacle lighting circuits just referred to, we are required to have a minimum of two small-appliance circuits in the kitchen, as explained in Chapter 2, and also a minimum of one 20-ampere laundry circuit. Remember, to these three minimum requirements for small-appliance and laundry circuits there shall be no permanently installed appliances connected, nor shall there be any lighting connected except that which shall be an integral part of the small appliances.

134

Referring to **Article 100** covering the definitions of appliances, we find three definitions—**fixed, portable, and stationary.** The intent of the small-appliance circuits is for use with portable appliances such as we use around the kitchen or in the family room. Refrigeration equipment may be included on the small appliance circuit according to **Section 220-3(b).** In my own home, I would consider that since both the refrigerator and the freezer contain valuable food items, that a separate circuit be run to each of these. This is not a Code requirement, but is purely a good practical suggestion. One could of course put a couple of lights or something of this sort on each of the additional circuits should he desire to do so. The same method would be true of the laundry circuit since this is for washing and ironing only and is not intended as a general purpose circuit, and should therefore have no lighting connected to it. One might wonder why a

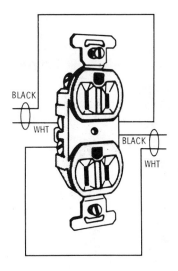

Fig. 74. A two-wire branch-circuit connection at a receptacle.

20-ampere circuit is necessary for a laundry. Many washers pull approximately 20 amperes when starting the spin cycle.

Going back to the minimum of two small-appliance circuits in the kitchen (covered by **Section 220-3(b)** of the NEC) anyone who understands the why and wherefore of installing these circuits, and what they are used for, will certainly add more than the two required. The NEC tells us that there is to be a minimum of two circuits in the kitchen, and it also tells us that there is to be a 20-ampere small-appliance circuit in the dining room, family room, breakfast room, and pantry. The question now comes up as to what is a family room? There is no practical definition listed, so we look to the room usage to define it. The intent, I am sure, is to have a small-appliance circuit available in a room where the family gathers to enjoy themselves and to have snacks which will require the use of portable electrical appliances.

There is nothing in the NEC which prohibits us from extending one or both of the small-appliance circuits to one or more of the rooms other than the kitchen which are required to have small-appliance circuits in them. But again be reminded that, for the small dollar investment required at the time of home construction, it is recommended that more than just the two small-appliance circuits be installed.

The small-appliance circuits in the kitchen and elsewhere may be run as a 2-wire, 115-volt circuit for each run, and this will provide a duplex receptacle at the same location, and each will be on a single 115-volt circuit. In this case, should a toaster and a fry pan both be plugged into this receptacle at the same time, the circuit breaker will probably trip. We have another alternative—to run a 3-wire multiwire circuit. In this type of circuit, the black conductor connects to phase A, the red conductor to phase B, and the white conductor to the neutral. This circuit may be run around the perimeter of the kitchen area. Receptacles come with a breakoff piece on the hot or brass-colored side of the terminals, and the black conductor connects to one brass screw, and the white conductor to either of the silver-colored screws. This will provide two circuits at each receptacle. Refer to Figs. 74 and 75, which illustrate this procedure.

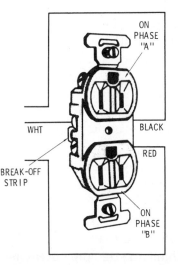

Fig. 75. Connecting a receptacle to provide two separate circuits.

The multiwire circuit does have one drawback which might cause an accident in a very rare case—should two appliances which have a ground fault on the phase conductor to the appliance cases are both plugged into the same duplex receptacle, 230-volts might be present between the cases of the two appliances. This problem will fast disappear as we are required to have three-conductor cords—that is, a cord with two current-carrying conductors plus an equipment grounding conductor. The 1968 NEC took this into consideration in making a substantial start in this direction. Also,

137

ground fault interrupters are fast taking their place in the wiring field, and it is very possible that the use of these might become a requirement in many of our circuits as covered by the NEC. There is certainly nothing to prohibit the use of these ground fault interrupters, known as GFIs, in any wiring you might be doing now.

The NEC does not tell us how many small-appliance receptacles may be put on one small-appliance circuit. It tells us only the minimum number of circuits required. We must be practical in wiring a residence and not install an excess number of receptacles on one small-appliance circuit. A good guide line to follow is to have not more than three or possibly four receptacles on one circuit. Be advised, however, that this is not one of the minimum Code requirements.

In figuring the number of outlets for general illumination, be sure that each circuit does not supply more outlets than the 3 watts per square foot basis allows, in fact leave some spare capacity on each circuit for future use.

The NEC permits us to use 15-ampere circuits in these areas. You will also find that many regulations adopted by some cities prohibit the 15-ampere circuits for receptacles on general lighting circuits, requiring instead 20-ampere circuits. You will also find that one of the manufacturers of precut homes also supplies materials for 20-ampere receptacle circuits.

In figuring the wiring for general-purpose lighting circuitry, it is suggested that you evaluate the difference in cost of using conductors with 20-ampere ampacity instead of conductors with 15-ampere ampacity. Your lighting switches, receptacle, and boxes will be the same in either case, as will the labor for an NM cable installation. The only additional expenses will be the price differential between the two different sizes of NM cable, and you will find this to be of relatively small importance. However, a contractor bidding for the wiring on a dwelling might find that this differential in conductor prices would cause him to lose the contract. Therefore, the contractor must be practical and, in figuring his estimate, should figure it on the minimum basis and then give an alternate which includes the larger ampacity conductors. Many times an inadequate job of selling is done by the contractor when considering bids for wiring. As an example, when you walk into an appliance store to buy a refrigerator, where does the salesman automatically lead you? He takes you to the top of the line, and if he observes that you are not able to purchase the highest price refrigerator, he then goes to the next lower price refrigerator and so on until he reaches the model that he is able to convince you to buy. Why should we not do the same in the selling of wiring. It is not padding the customer's bill for the wiring, but it is giving him something a little better which he will soon find out was advantageous to him. Your best advertising will come from a satisfied customer.

Ground fault circuit interrupters are required on bathroom receptacles.

CHAPTER 13

Boxes and Fittings

Boxes and fittings are covered in **Article 370** of the NEC. This chapter will discuss the usage of such boxes, how they shall be mounted, and the number of conductors permitted in each. Notice that cubic inch capacity is the criteria for the number and sizes of conductors permitted in boxes. Since these boxes are not always easily measured to obtain the cubic content, they are sometimes filled with water and the cubic content of the water measured instead.

There is quite a variety of boxes used in residential wiring. Perhaps the most common is the outlet or switch box, often

Fig. 76. Various types of outlet or switch boxes.

referred to as a receptacle box. They are available in aluminum, galvanized steel, and plastic. Fig. 76 illustrates some of these boxes. Some are for use with conduit, and this type

Fig. 77. Various types of built-in cable clamps.

is available with or without plaster ears so that they may be mounted in either a plaster wall or in concrete and attached to rigid conduit or electrical metallic tubing. These same boxes may be used with armored cable or nonmetallic sheathed cable by using approved clamps to secure these cables in the box. They also are available with clamps mounted in the box, one type approved for armored cable and another type approved for NM cable. See Fig. 77. Be certain that the boxes used are approved for the type wiring being installed.

2-gang.

3-gang.

4-gang.

Fig. 78. Ganged boxes.

These boxes are available in single-gang types for mounting only one receptacle or switch, or for up to 3 switches and/or outlets of the DESPARD variety mounted in one single

142

yoke. They are also available in multigang types for 2 or more receptacles or switches mounted in one row horizontally. See Fig. 78. Practically all of the single-gang boxes are of the sectional variety with which, by removing one or both sides of two or more boxes, any size box can be assembled to allow the installation of multigang devices. Bakelite or plastic boxes must be purchased as ganged boxes in proper number since they cannot be ganged in the field. See Fig. 79.

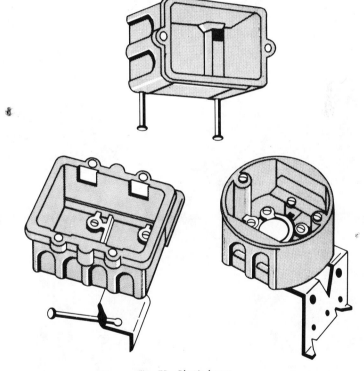

Fig. 79. Plastic boxes.

In using metal boxes it is required that the cables be stapled within 12 inches of where they enter the box, and

143

Fig. 80. Various types of nail-on boxes.

built-in clamps or approved connectors shall be used. Plastic boxes do not have built-in clamps for the cables, so it therefore is necessary to staple the cables within 8 inches of the box. In addition, NM cable is the only type permitted to be used with plastic boxes.

Boxes are available with plaster ears for mounting to wooden lath, but the majority are what are called nail-on boxes, as shown in Fig. 80. Through-type boxes for mounting switches or receptacles on both sides of a wall are also manufactured. These are shown in Fig. 81.

Fig. 81. A through-type box for installing switches or receptacles on both sides of a wall.

Extensions are available which may be mounted on boxes to give a larger cubic capacity, or for extending the box if it is mounted too far back, or for use if at some future date it is necessary to furr the wall out for panelling. See Fig. 82.

Where lath and plaster is used, the boxes usually become filled with plaster and hard to clean out and are also

145

Fig. 82. An extension for a switch box.

sometimes hard to find. Threads must be cleaned after plastering. *Appleton Electric Products* manufactures a cover for protection of boxes in such installations. See Fig. 83. This cover has a piece of gummed paper in the center with the notation, "Remove Tape Before Plastering." The box is

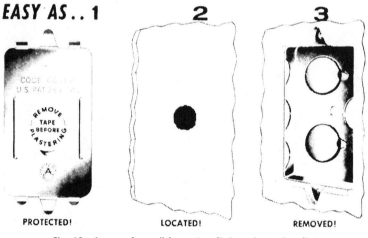

Fig. 83. A cover for wall boxes installed in plastered walls.

plastered over, but a small blue spot bleeds through the plaster and marks the position of the outlet. The plaster may then be removed, leaving a clean box that is well plastered in. Receptacle and switch boxes shall be mounted flush with all combustible materials, but may be set back ¼ inch when installed in noncombustible materials. In all cases, however, the wall material must fit closely around the boxes.

For mounting switches and receptacles, what is commonly known as a 4" square box (Fig. 84) is often used. Fig. 84A shows a plain conduit type, but all of these boxes are available with clamps for NM cable or armored cable. Figs. 84B, C, and D illustrate 4" square boxes which may be attached to wooden studs without the use of nails. In the event a 4" square box does not have enough capacity for the number

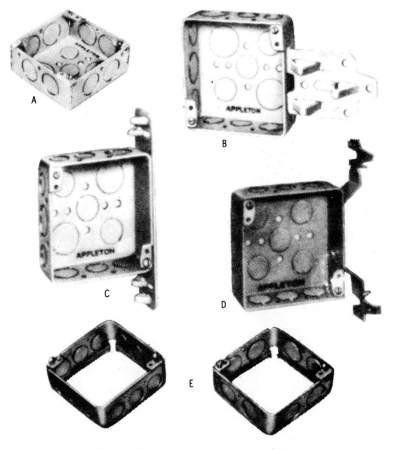

Fig. 84. Various types of 4-inch square boxes.

and sizes of conductors that are being installed, extensions such as those shown in Fig. 84E may be added. One or more extensions may be used as required to obtain sufficient capacity. Note that 4″ square boxes are not equipped for

Fig. 85. Plaster rings for 4-inch square boxes.

mounting receptacles and/or switches, so it is necessary to use what is known as a plaster ring (Fig. 85). Although the word "plaster ring" is used, it does not mean that they are only to be used on plastered walls, since they are also usable for panelling or drywalls. They are available in dif-

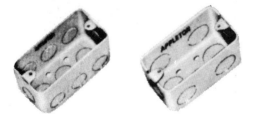

Fig. 86. Handy boxes.

ferent depths to accommodate the type wall in which they are being installed.

148

Another type of box, called a "handy box," is shown in Fig. 86. These may be purchased with nail-on adapters, but are usually used for surface mounting. They do not come with clamps, so are for use with conduit or EMT. If it is desired to use them with armored cable or NM cable (Ro-MEX), approved clamps must be used, such as those that

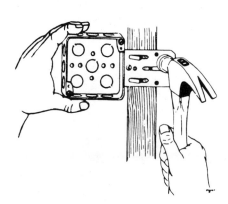

Fig. 87. An extension for a handy box.

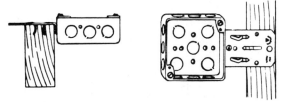

Fig. 88. Method of installing self-nailing square boxes.

were shown in Fig. 76. Extensions are also available for handy boxes (Fig. 87) to increase their capacity.

149

Fig. 88 shows the method used to install a self-nailing 4″ square box. The same method is used for self-nailing handy boxes and regular type receptacle boxes.

Fig. 89. A plain 4-inch octagonal box.

Fig. 89 illustrates a 4″ octagon box of the conduit or EMT type. This kind of box may also be used with cables, provided that approved clamps are added, or that the box has approved built-in clamps (Fig. 90). Fig. 91 shows 4″ octa-

Fig. 90. Four-inch octagonal boxes with built-in cable clamps.

gon boxes of the nail-on type. Extension rings (Fig. 92) are available for 4″ octagon boxes to gain sufficient capacity for the number and sizes of conductors being used.

If the walls or ceilings are plastered, or if drywall is used, 4″ octagon boxes should have plaster rings of a type similar to those shown in Fig. 85. This type of box may be mounted with mounting brackets similar to those shown for a common box or a 4″ square box. It is also possible to install a 2″ x 4″ between the ceiling joists and mount the octagon box on

150

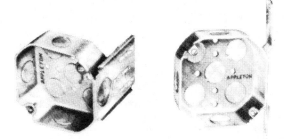

Fig. 91. Nail-on type octagonal boxes.

Fig. 92. Extension rings for octagonal boxes.

this. There are, however, many and varied types of patented hangers available, two of which are shown in Fig. 93.

Fig. 94 shows a plaster ring installed on a 4″ square box. Remember, these rings are available in various depths, so purchase the one required.

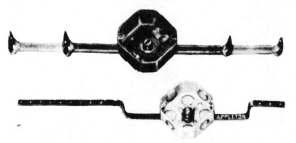

Fig. 93. Mounting brackets for octagonal boxes.

151

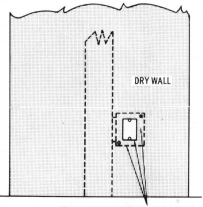

4" SQUARE BOX WITH PLASTIC RING AND DRYWALL CUT OUT
TO FIT THE RECEPTACLE PORTION OF THE PLASTER RING

Fig. 94. A plaster ring installed on a 4-inch square box.

A very vital portion of **Article 370** is **Section 370-6** concerning the number of conductors allowed in a box. This is of such vital importance that **Section 370-6** and **Table 370-6(a)** and **370-6(b)** are reproduced here along with any necessary comments.

370-6. Number of Conductors in Switch, Outlet, Receptacle, Device, and Junction Boxes. *Boxes shall be of sufficient size to provide free space for all conductors enclosed in the box.*

The provisions of this **Section** *shall not apply to terminal housings supplied with motors. (See* **Section 430-12**).

(a) **Standard Boxes.** *The maximum number of conductors, not counting fixture wires permitted in standard boxes, shall be as is listed in* **Table 370-6(a)**. *See* **Section 370-18** *where boxes or conduit bodies are used as junction or pull boxes.*

(1) **Tables 370-6(a)** *shall apply where no fittings or devices such as fixture studs, cable clamps, hickeys,*

switches, or receptacles, are contained in the box and where no grounding conductors are part of the wiring within the box. Where one or more of these types of devices, such as fixture studs, cable clamps, or hickeys, are contained in the box, the number of conductors shown in the table shall be reduced by one for each type of device; an additional deduction of one conductor shall be made for each strap containing one or more devices; and a further deduction of one conductor shall be made for one or more grounding conductors entering the box. A conductor running through the box shall be counted as one conductor, and each conductor originating outside the box and terminating inside the box is counted as one conductor. Conductors, no part of which leaves the box, shall not be counted. The volume of a wiring enclosure (box) shall be the total volume of the assembled sections, and where used, the space provided by plaster rings, domed covers, extension rings, etc. that are marked with their volume in cubic inches or are made from boxes the dimensions of which are listed in **Table 370-6(a)**.

(2) For combinations of conductor sizes shown in **Table 360-6(a)**, the volume per conductor listed in **Table 370-6(b)** shall apply. The maximum number and size of conductors listed in **Table 370-6(a)** shall not be exceeded.

(b) **Other Boxes.** Boxes 100 cubic inches or less other than those described in **Table 370-6(a)** and conduit bodies having provisions for more than two conduit entries shall be durably and legibly marked by the manufacturer with their cubic inch capacity and the maximum number of conductors permitted shall be computed using the volume per conductor listed in **Table 370-6(b)** and the deductions provided in **Section 370-6(a)(1)**. Boxes described in **Table 370-6(a)** that have a larger cubic inch capacity than is designated in the Table shall be permitted to have their cubic inch capacity marked as required by this Section and

153

the maximum number of conductors permitted shall be computed using the volume per conductor listed in **Table 370-6(b)**.

Where **#6** conductors are installed the minimum wire bending space required in **Table 373-6(a)** shall be provided.

(c) **Conduit Bodies.** Conduit bodies enclosing **#6** conductors or smaller shall have a cross-sectional area not less than twice the cross-sectional area of the largest conduit to which it is attached. The maximum number of conductors permitted shall be the maximum number permitted in **Table 1**, Chapter 9, for the conduit to which it is attached.

Table 370-6(a). Metal Boxes

Box Dimension, Inches Trade Size or Type	Min. Cu. In. Cap.	Maximum Number of Conductors				
		#14	#12	#10	#8	#6
4 x 1¼ Round or Octagonal	12.5	6	5	5	4	0
4 x 1½ Round or Octagonal	15.5	7	6	6	5	0
4 x 2⅛ Round or Octagonal	21.5	10	9	8	7	0
4 x 1¼ Square	18.0	9	8	7	6	0
4 x 1½ Square	21.0	10	9	8	7	0
4 x 2⅛ Square	30.3	15	13	12	10	6*
4¹¹⁄₁₆ x 1¼ Square	25.5	12	11	10	8	0
4¹¹⁄₁₆ x 1½ Square	29.5	14	13	11	9	0
4¹¹⁄₁₆ x 2⅛ Square	42.0	21	18	16	14	6
3 x 2 x 1½ Device	7.5	3	3	3	2	0
3 x 2 x 2 Device	10.0	5	4	4	3	0
3 x 2 x 2¼ Device	10.5	5	4	4	3	0
3 x 2 x 2½ Device	12.5	6	5	5	4	0
3 x 2 x 2¾ Device	14.0	7	6	5	4	0
3 x 2 x 3½ Device	18.0	9	8	7	6	0
4 x 2⅛ x 1½ Device	10.3	5	4	4	3	0
4 x 2⅛ x 1⅞ Device	13.0	6	5	5	4	0
4 x 2⅛ x 2⅛ Device	14.5	7	6	5	4	0
3¾ x 2 x 2½ Masonry Box/gang	14.0	7	6	5	4	0
3¾ x 2 x 3½ Masonry Box/gang	21.0	10	9	8	7	0
FS — Minimum Internal Depth 1¾ Single Cover/Gang	13.5	6	6	5	4	0
FD — Minimum Internal Depth 2⅜ Single Cover/Gang	18.0	9	8	7	6	3

Conduit bodies having provisions for less than three conduit entries shall not contain splices, taps, or devices, unless they comply with the provisions of 370-6(b) and are supported in a rigid and secure manner.

Table 370-6(a). Boxes

Box Dimension, Inches Trade Size or Type	Min. Cu. In. Cap.	Maximum Number of Conductors				
		#14	#12	#10	#8	#6
FS — Minimum Internal Depth 1¾ Multiple Cover/Gang	18.0	9	8	7	6	0
FD — Minimum Internal Depth 2⅜ Multiple Cover/Gang	24.0	12	10	9	8	4

*Not to be used as a pull box. For termination only.

Table 370-6(b). Volume Required Per Conductor

Size of Conductor	Free Space Within Box for Each Conductor
No. 14	2. cubic inches
No. 12	2.25 cubic inches
No. 10	2.5 cubic inches
No. 8	3. cubic inches
No. 6	5. cubic inches

370-6. Number of Conductors in a Box—*Boxes shall be of sufficient size to provide free space for all conductors enclosed in the box.* This is a basic and broad statement which, in itself, is quite sufficient; however, a complete analysis of this will be made as it is in the Code. The main point is that in the installation of conductors in boxes, the conductors should never be forced into the box as this is a potential source of trouble. The Code spells out what is good practice as well as the minimum requirements. The installer should, however, always bear in mind the intent and, if necessary to do a good job, go even further than the minimum requirements.

The limitations imposed by **Sections 370-6 (a** and **b)** are not intended to apply to terminal housings supplied with

155

motors, nor to those types of boxes or fittings without knock-outs that have hubs or recessed parts for terminal bushings and locknuts.

(a) **Table 370-6(a)** cover the maximum number of conductors that will be permitted in outlet and junction boxes. There has been no allowance in these Tables for fittings or devices, such as studs, cable clamps, hickeys, switches, or receptacles that are contained in the box. These must be taken into consideration and deductions made for them. A deduction of one conductor shall be made for each of the following:

(1) One or more fixture studs.

(2) Cable clamps.

(3) Hickeys.

(4) Grounding conductors count as one.

There shall be a further deduction of one conductor for one or several flush devices mounted on the same strap. A conductor that runs through a box is counted as only one conductor. A conductor originating out of the box and terminating in the box is counted as one conductor. Conductors of which no part leaves the box will not be counted, such as wires to fixtures that are spliced onto the other wires in the box.

Boxes are often ganged together with more than one device per strap mounted in these ganged boxes. In these cases, the same limitations will apply as if they were individual boxes.

Please refer to Figs. 95 through 98 for illustrations concerning this Section.

Tables 370-6 will not be applicable. In such cases refer to **Table 360-6 (b)** from which the number of conductors that will be permitted in the box can be calculated. The various wire sizes and the space that will be allowed per conductor

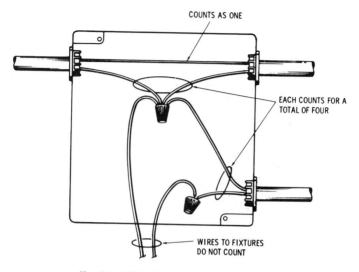

COUNTS AS ONE

EACH COUNTS FOR A
TOTAL OF FOUR

WIRES TO FIXTURES
DO NOT COUNT

Fig. 95. Which wires to count in a junction box.

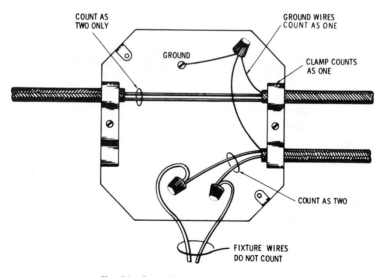

COUNT AS
TWO ONLY

GROUND

GROUND WIRES
COUNT AS ONE

CLAMP COUNTS
AS ONE

COUNT AS TWO

FIXTURE WIRES
DO NOT COUNT

Fig. 96. Grounding wire counts as one.

157

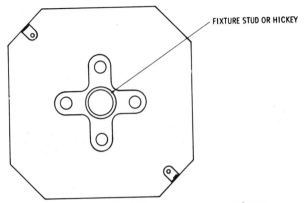

Fig. 97. Fixture stud or hickey counts as one conductor.

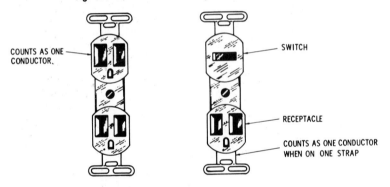

Fig. 98. How to count devices in figuring fill.

are given. It is only necessary to calculate the cubic space in the box, and multiply by the free space within the box for each conductor. If the calculation exceeds the cubic size of the box, the next larger box will have to be used or an extension added.

The question concerning plaster rings and the effect that they have on the conductors permitted in the box is always arising. This is covered in the NEC. Although the plaster ring is meant for another purpose, almost every inspector will agree that extra space is provided by the plaster ring and, based on the cubic inches that it adds, he requires allowances for this by use of a larger box in case more room be needed. The entire intent is not to crowd the conductors and cause failure of the insulation in so doing.

370-7. Conductors Entering Boxes or Fittings—Care shall be exercised in protection of the conductors from abrasion where they enter the boxes or fittings. With conduit, this is

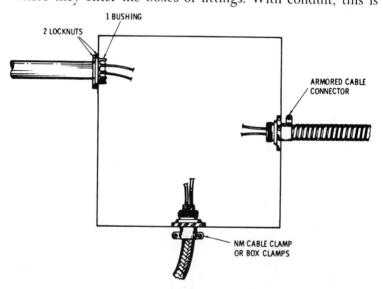

Fig. 99. Connection of cables and conduit to boxes.

accomplished by bushings or other approved devices. With NM cable, as may be seen in Fig. 99, the outer covering of the cable should protrude from the clamp to provide this protection. With armored cable, fiber bushings are to be inserted between the conductors and the armor to prevent any abrasion. The following shall be complied with:

(a) **Openings to Be Closed.** *Openings through which conductors enter shall be adequately closed.* Where

159

single conductors enter the boxes, loom covering is to be provided; with cable, cable clamps shall be used or the boxes provided with built-in cable clamps; with conduit, the locknuts and bushings will adequately close the openings.

(b) **Metal Boxes and Fittings.** When metal boxes or fittings are used with open wiring, proper bushings

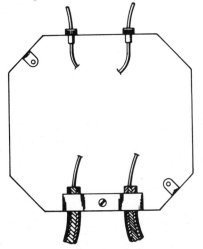

Fig. 100. Open wiring into boxes.

shall be used. In dry places, a flexible tubing may be used and extended from the last conductor support into the box and secured. See Fig. 100.

Where a raceway or cable enters the box or fitting, the raceway or cable is to be properly secured to the box or fitting. With conduit, two locknuts and a bushing should be used. With armored cable, approved connectors shall be used. With NM cable, a connector or built-in clamps shall be used.

(c) **Nonmetallic Boxes.** Where nonmetallic boxes are used with either concealed knob-and-tube work or open wiring, the conductors shall pass through individual holes in the box. If flexible tubing is used over the conductor, it shall extend from the last conductor support into the hole in the box.

Where nonmetallic cable is used, it shall extend through the opening in the box. It is not required that individual conductors or cables be clamped if the individual conductors or cables are supported within 8 inches of the box. When nonmetallic conduit is used with nonmetallic boxes, the conduit shall be connected to the box by approved means.

CHAPTER 14

Spacing of Receptacles

The spacing of receptacles in a residential occupancy is not very complicated once it is understood what is meant to be accomplished. This is very thoroughly covered in **Section 210-52** of the NEC.

Dwelling Unit Receptacle Outlets

(a) **General Provisions.** *In every kitchen, family room, dining room, breakfast room, living room, parlor, library, den, sun room, bedroom, recreation room, or similar rooms, receptacle outlets shall be installed so that no point along the floor line in any wall space is more than 6 feet, measured horizontally, from on outlet in that space, including any wall space 2 feet or more in width and the wall space occupied by sliding panels in exterior walls. The wall space afforded by fixed room dividers, such as free-standing bar-type counters, shall be included in the 6-foot measurement.*

As used in this Section a "wall space" shall be considered a wall unbroken along the floor line by doorways, fireplaces, and similar openings. Each wall space 2 or more feet (610 mm or more) wide shall be treated individually and separately from the other wall spaces within the room. A wall space shall be permitted to include two or more walls of a room (around corners) where unbroken at the floor line.

The purpose of this requirement is to minimize the use of cords across doorways, fireplaces, and similar openings.

Receptacle outlets shall, insofar as practicable, be spaced equal distances apart. Receptacle outlets in floors shall not be counted as part of the required number of receptacle outlets unless located close to the wall.

The receptacle outlets required by this Section shall be in addition to any receptacle that is part of any lighting fixture or appliance, located within cabinets or cupboards, or located over 5½ feet (168 m) above the floor.

Exception: Permanently installed electric baseboard heaters equipped with factory-installed receptacle outlets or outlets provided as a separate assembly by the manufacturer shall be permitted as the required outlet or outlets for the wall space utilized by such permanently installed heaters. Such receptacle outlets shall not be connected to the heater circuits.

(b) **Counter Tops.** *In kitchen and dining areas a receptacle outlet shall be installed at each counter space wider than 12 inches. Counter top spaces separated by range tops, refrigerators, or sinks shall be considered as separate counter top spaces. Receptacles rendered inaccessible by the installation of stationary appliances shall not be considered as these required outlets.*

(c) **Bathrooms.** *In dwelling units, at least one wall receptacle outlet shall be installed in the bathroom adjacent to the basin location. See* **Section 210-8(a)(3).**

This receptacle shall have a ground-fault circuit-interrupter.

(d) **Outdoor Outlets.** *For one- and two-family dwellings, at least one receptacle shall be installed outdoors. See* **Section 210-8(a)(3).**

This receptacle outlet shall have a ground-fault circuit-interrupter.

(e) **Laundry Area.** *In dwelling units, at least one receptacle outlet shall be installed for the laundry.*

(f) **Basement and Garages.** *For a one-family dwelling, at least one receptacle outlet in addition to any provided for laundry equipment shall be installed in each basement and in each attached garage. See* **Section 210-8(a)(2).**

163

This receptacle outlet in the garage shall have a ground-fault circuit-interrupter.

Exception No. 1: In a dwelling unit that is an apartment or living area in a multifamily building where laundry facilities are provided on the premises that are available to all building occupants, a laundry receptacle shall not be required.

Exception No. 2: In other than single-family dwellings where laundry facilities are not to be installed or permitted, a laundry receptacle shall not be required.

As used in this Section a "wall space" shall be considered a wall unbroken along the floor line by doorways, fireplaces, and similar openings. Each wall space two or more feet wide shall be treated individually and separately from other wall spaces within the room. A wall space shall be permitted to include two or more walls of a room (around corners) where unbroken at the floor line.

The purpose of this requirement is to minimize the use of cords across doorways, fireplaces, and similar openings.

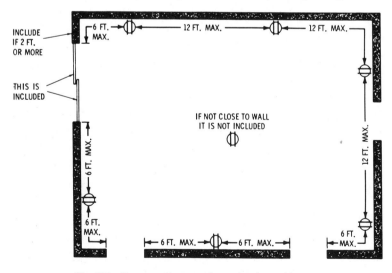

Fig. 101. Proper wall receptacle spacing for residences.

The receptacle outlets required by this Section shall be in addition to any receptacle that is part of any lighting fixture or appliance, located within cabinets or cupboards, or located over 5½ feet above the floor.

This portion of the Code is concerned with the fact that, in most all cases, any electrical device to be plugged in shall be no more than 6 feet from any receptacle. See Fig. 101.

In bedrooms, for example, where there are closets with sliding doors, these doors do not count as wall space, but receptacles must be installed no more than 6 feet on either side of sliding doors.

In laundry rooms, or where special appliances will be used, a receptacle is not to be more than 6 feet from the location where the laundry equipment or special appliance is to be plugged in. You might ask why so many receptacles? These are really not an excessive number of receptacles if you notice that appliances with cords attached have cords about 6 feet long. Thus, it is unnecessary to have long extension cords which can be a definite fire hazard when they are overloaded (which they often are). Many owners will ask for more receptacles than are required as a minimum by the NEC, and certainly it is your obligation to install as many receptacles over the minimum as they desire.

The height for installing switches and receptacles in a residence will vary with the desires of the owner or builder, and their desires should be followed. However, the normal height above the floor is 12″ to the center of a receptacle. Where kitchen receptacles are above counters (and most counters in kitchens have a height of 3 feet) the receptacle must be mounted above this height, so 42″ to the center of the receptacle is considered as a standard height.

A common height for switches is 54″ to their centers, and again this will vary with the location and desire of the owner, but this is a practical height for a rule-of-thumb location.

CHAPTER 15

Mobile Homes

A mobile home is actually a residence, but because of its mobility the structure has to be different from a regular residential dwelling. The definition in **Section 550-2** defines a mobile home as:

A factory-assembled structure or structures equipped with the necessary service connections and made so as to be readily moveable as a unit or units on its own running gear and designed to be used as a dwelling unit(s) without a permanent foundation.

> *The phrase "without a permanent foundation" indicates that the support system is constructed with the intent that the mobile home placed thereon will be moved from time to time at the convenience of the owner.*

The nature of the construction of a mobile home requires some deviations from the wiring of a permanent residence. From the NEC definition, it would appear that when mounted on a foundation, this home would then cease to be a mobile home, and it is entirely possible that inspection authorities might classify it as a modular or prefabricated home and subject to the same wiring requirements and regulations as any other residence.

Due to its portability, and its being mounted on a chassis and running gear, there are requirements which must be different from those for wiring in a permanent residence. This is not to imply that safety can be overlooked, especially since many mobile homes have metal siding which might become electrically energized. It might be stated that extra safety precautions should be taken in wiring a mobile home. Also, the fact that it has to be transported on wheels to a site location must not be overlooked.

Underwriter's Laboratories, Inc., has a labeling and inspection service for mobile homes which, I am sure, would be acceptable to most inspection authorities. Where the homes are not UL approved and labeled, then the local inspection authority in many cases will want to inspect the wiring during and after completion.

Some of the differences between the wiring of mobile homes and the standard residence will be covered here. The subject will not be covered in its entirety, so reference to the current National Electrical Code will be necessary for specific details. Points not covered in **Article 550** of the NEC will be subject to the provisions of the NEC as covered in the wiring of a regular residence.

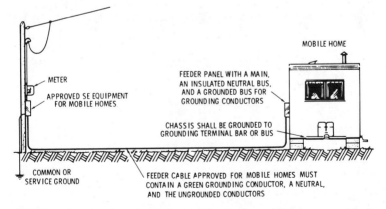

Fig. 103. Pole-mounted service-entrance equipment for mobile home use with the feeder cable above ground.

The mobile home service equipment shall be located adjacent to the mobile home, and not in or on the mobile home. The power supply to the mobile home will be feeder circuits

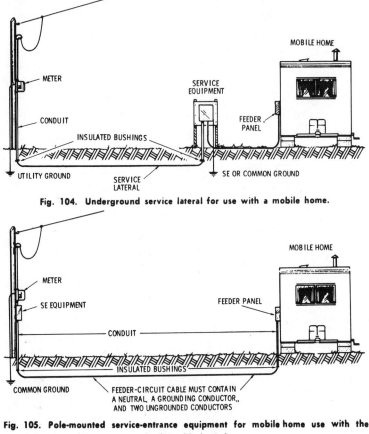

Fig. 104. Underground service lateral for use with a mobile home.

Fig. 105. Pole-mounted service-entrance equipment for mobile home use with the feeder cable buried.

consisting of not more than three 50-ampere mobile home supply cords, or by a permanently installed circuit.

Refer to Figs. 103, 104, 105, and 106 for various methods of installing the service and feeders to the mobile home.

Please note the feeder panel is shown mounted on the exterior of the mobile home. This is merely to clarify the intent of the drawings. Usually it is mounted in the interior.

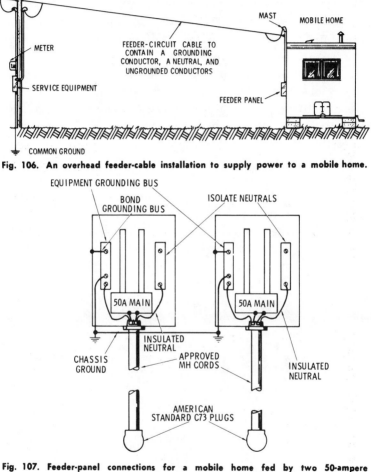

Fig. 106. An overhead feeder-cable installation to supply power to a mobile home.

Fig. 107. Feeder-panel connections for a mobile home fed by two 50-ampere supply cords.

Each feeder panel fed by a cord shall not be electrically interconnected to other feeder panels fed by a cord. That is, these shall not be electrically interconnected on either the line or load sides, except that the grounding circuits and grounding means shall be electrically interconnected. Fig. 107 illustrates a mobile home fed by two 50-ampere cords. Three cords were not used in order to simplify the illustration, but the same conditions would also apply if a third supply cord where to be used.

If cord feeders are not used, but a permanent connection is made instead, such as illustrated in Figs. 105 and 106, one feeder could supply and one panel would suffice. All panels are to have main disconnects with overcurrent protection.

Range and dryer frames are not permitted to be connected to the neutral as allowed in **Section 250-60**, but are required to have a fourth equipment grounding conductor supplied, which may be bare or green insulated and connected to the frame, with the neutral isolated and the bonding terminal usually supplied with ranges and dryers to be discarded.

Mobile home supply cords shall not be more than 36½ feet in length, nor less than 21 feet.

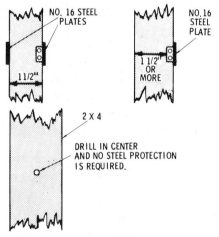

Fig. 108. Methods of protecting mobile home wiring.

Because studs as small as 1½″ are often used in mobile home construction, 16 gauge steel protection will be required to protect the wiring from nails, staples, etc. NM cable is usually used in the wiring of these homes, and could be very easily penetrated by a nail or staple on these small studs. If standard 2″ x 4″ studs are used, and are drilled in the center, this steel protection will not be required. See Fig. 108.

One very important item is the receptacle and switch box sizes. Because small joists are often used, it is hard in many cases to get boxes with the cubic content necessary to meet the fill requirements of **Article 370.** This Article must either be followed or boxes approved for the purpose used.

Section 550-9 concerning grounding will be quoted from the 1981 NEC.

Section 550-9. Grounding.

Grounding of both electrical and nonelectrical metal parts in a mobile home is through connection to a grounding bus in the mobile home distribution panel. The grounding bus is grounded through the green-colored conductor in the supply cord or the feeder wiring to the service ground in the service-entrance equipment located adjacent to the mobile home location. Neither the frame of the mobile home nor the frame of any appliances may be connected to the neutral conductor in the mobile home.

(a) Insulated Neutral.

(1) The grounded circuit conductor (neutral) shall be insulated from the grounding conductors and from the equipment enclosures and other grounded parts. The grounded (neutral) circuit terminals in the distribution panel and in ranges, clothes dryers, counter-mounted cooking units and wall-mounted ovens shall be insulated from the equipment enclosure. Bonding screws, straps, or buses in the dis-

171

tribution panel or in appliances are to be removed and dis-carded.

(2) Connection of ranges and clothes dryers with 115/230-volt, 3-wire ratings shall be made with 4 conductor cord and 3-pole, 4-wire grounding-type plugs, or by Type AC cable or conductors enclosed in flexible metal conduit.

(b) Equipment Grounding Means.

(1) The green-colored insulated grounding wire in the supply cord or permanent feeder wiring shall be connected to the grounding bus in the distribution panel or disconnect-ing means.

(2) In the electrical system, all exposed metal parts, en-closures, frames, lamp fixture canopies, etc., shall be effec-tively bonded to the grounding terminal or enclosure of the distribution panel.

(3) Cord-connected appliances, such as washing ma-chines, clothes dryers, refrigerators, and the electrical sys-tem of gas ranges, etc., shall be grounded by means of a cord with grounding conductor and grounding type attach-ment plug.

(c) Bonding of Noncurrent-Carrying Metal Parts.

(1) All exposed noncurrent-carrying metal parts that may become energized shall be effectively bonded to the ground-ing terminal or enclosure of the distribution panelboard. A bonding conductor shall be connected between the distri-bution panelboard and an accessible terminal on the chassis.

(2) Grounding terminals shall be of the solderless type and approved as pressure-terminal connectors recognized for the wire size used. The bonding conductor shall be solid or stranded, insulated or bare, and shall be No. 8 copper minimum, or equal. The bonding conductor shall be routed so as not to be exposed to physical damage.

(3) *Metallic gas, water, and waste pipes and metallic air circulating ducts shall be considered bonded if they are connected to the terminal on the chassis (see* **Section 550-9)** (c)(1)) *by clamps, solderless connectors, or by suitable grounding-type straps.*

(4) *Any metallic roof and exterior covering shall be considered bonded if (*a*) the metal panels overlap one another and are securely attached to the wood or metal frame parts by metallic fasteners, and (*b*) if the lower panel of the metallic exterior covering is secured by metallic fasteners at a cross member of the chassis by two metal straps per mobile home unit or section at opposite ends.*

The bonding strap material shall be a minimum of 4 inches in width of material equivalent to the skin or a material of equal or better electrical conductivity. The straps shall be fastened with paint-penetrating fittings, such as screws and starwashers or equivalent.

550-10. Testing.

(a) **Dielectric Strength Test.** *The wiring of each mobile home shall be subjected to a 1-minute, 900-volt, dielectric strength test (with all switches closed) between live parts (including neutral) and the mobile home ground. Alternatively, the test shall be permitted to be performed at 1,080 volts for 1 second. This test shall be performed after branch circuits are complete and after fixtures or appliances are installed.*

Exception: Fixtures or appliances which are approved shall not be required to withstand the dielectric strength test.

Another important part of the Code concerning mobile homes is **Section 550-10.** This concerns testing. It is highly recommended that this test be run at the completion of the wiring of each mobile home, and it is further recommended that it also be run when the mobile home is delivered to the site and also after each moving. Notice that this is not a Code requirement (to test at each site), but it would certainly be a step toward better safety.

Each receptacle should be checked for proper polarity and equipment ground continuity. In checking mobile wiring, it was found that the majority of rejects were due to improper grounding or polarities. Rechecks of inspections are time consuming and expensive to both the manufacturer and the inspection authority.

CHAPTER 16

Wiring For Electric House Heating

Electric heating is fast being accepted as the method of heating homes. Since this is the case, this subject should be covered. The secret of good economical house heating by means of electricity lies in following certain tried and true methods of installing insulation and vapor barriers in an accepted manner. The figuring of a house heating system is another subject—if you are not fully qualified on this subject consult your local utility company since they have qualified engineers for advice and calculations. They are interested in low-cost heating and will lay out the job on the most economical basis. It has been known that (in the past) some utility companies have supplied the materials at slightly below cost in order to increase their consumer listing.

When I started in the utility part of the electrical field, the peak loads came mostly in December. This has changed and summer loads have, in most cases, become the peak loads due to irrigation and air conditioning. Thus, they are interested in adding winter loads. Feel free to call on them.

The insulation that you will use is vital. Many electrical contractors that install heating also install insulation and in doing so save themselves many problems that might arise due to an improperly insulated home. This would cause high electric bills, which are entirely unnecessary. The insulating factor and the thickness of the insulation play an important part in the cost of heating. With a well insulated home, the humidity will rise which adds to their home comfort in the winter. With the higher humidity, a vapor-barrier must be installed between the insulation and the drywall or lathe and plaster, as moisture will cut the efficiency of the insulation.

It might be well to consider a humidistat control system to keep the humidity at the proper level. This would be especially true in damp climates. The major heat losses are through windows, doors, and ceilings, so attention must be made to these items. The construction of the house has no place in this book, but is mentioned in order that you might fulfill your job well.

There are many types of electric heating systems, such as:

1. Ceiling heat cables,
2. Central heating systems,
3. Baseboard heating, which may be conventional resistive heating units, or baseboard water-type heating units,
4. Heating panels,
5. Unit heaters,
6. Heat pumps.

A portion of the material used here will be taken from *Audel's Guide to the 1981 National Electric Code,* which in an interpretation of the NATIONAL ELECTRICAL CODE requirements.

For the sizing of overcurrent devices and branch-circuit conductors for electrical space-heating (fixed) they should be calculated on the basis of 125% of the total load of the heaters and motors (if equipped with motors) and 125% of the total load of the heaters, if not equipped with motors.

Example: If a heating circuit draws 14-amperes full load, the circuit conductors and the overcurrent devices will have to be sized to a minimum of 125% of 14-amperes, or 17½-amperes. So a 20-ampere breaker or fuses will be required and No. 12 copper conductors or No. 10

aluminum or copper-clad aluminum conductors will be required.

Usually each room has its own thermostatic control. Some thermostats have an "OFF" position and some do not. Where there is an "OFF" position, the thermostat switch must break all hot conductors (not the neutral) simultaneously. See Fig. 109.

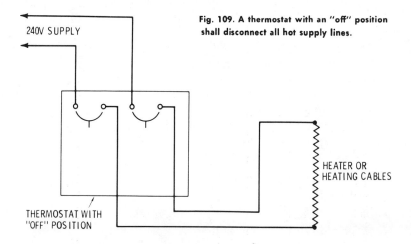

240V SUPPLY

Fig. 109. A thermostat with an "off" position shall disconnect all hot supply lines.

HEATER OR HEATING CABLES

THERMOSTAT WITH "OFF" POSITION

A very common method of heating homes is by means of ceiling heat cable. This is a very effective means of heating (and quite trouble free) if installed properly. One might ask: *What happens if a cable opens after the installation has been in operation for some time?* There are "open cable" locating equipment available and the ceiling may be opened at that point and repairs made. But, be sure that the proper equipment is used in making the repairs. Fig. 110 illustrates ceiling heat cable construction. Note the 7 ft. of nonheating leads. These SHALL NOT BE SHORTENED—the ends have the markings required by the NEC and UL laboratories. The heat cable may be identified by

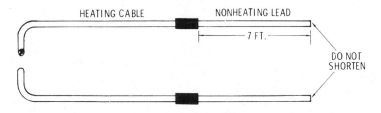

Fig. 110. Nonheating leads shall be a minimum of 7 feet.

the markings at the ends of the nonheating leads and by the color. Be sure the cables used carry the UL listing mark:

120- volt nominal voltage ....Yellow
208- volt nominal voltage ....Blue
240- volt nominal voltage ....Red

There are other colors for other voltages, but these are the voltages which you find in homes. See the *Audel's Guide to the 1981 NEC.*

424-38. Area Restrictions.

(a) **Shall Not Extend Beyond the Room or Area.** *Heating cables shall not extend beyond the room or area in which they originate.* See Fig. 111.

(b) **Uses Prohibited.** *Cables and panels shall not be installed in closets, over walls or partitions that extend to the ceiling, or over cabinets whose clearance from the ceiling is less than the minimum horizontal dimension of the cabinet to the nearest cabinet edge that is open to the room or area.*

Exception: Isolated runs of cable may pass over partitions where they are embedded.

(c) **In Closet Ceilings as Low Temperature Heat Source to Control Relative Humidity.** There are climates where humidity control is required in closets, but the prohibition of cables in closets is not intended to prohibit the use of low-temperature humidity controls in closets. See Fig. 111.

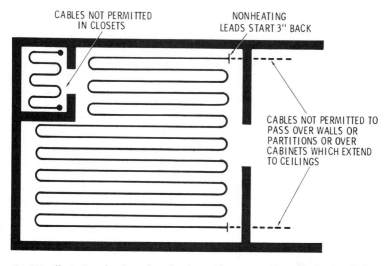

CABLES NOT PERMITTED
IN CLOSETS

NONHEATING
LEADS START 3" BACK

CABLES NOT PERMITTED TO
PASS OVER WALLS OR
PARTITIONS OR OVER
CABINETS WHICH EXTEND
TO CEILINGS

Fig. 111. Illustration showing where heating cables may and may not be installed in the ceilings.

424-39. Clearance from Other Objects and Openings— *Panels and cables shall be separated at least 8 inches from the edge of outlet boxes and junction boxes that are to be used for mounting surface lighting fixtures and their trims, ventilation openings and other such openings in room surfaces. Sufficient area shall be provided to assure that no heating cable or panel will be covered by other surface mounted lighting units.* The temperature limits and overheating of the cables are involved in this instance. Therefore, the requirements of this Section should be very carefully followed and, if in doubt, a little extra clearance should be given. See Fig. 112.

424-40. Splices—Splicing of cables is prohibited except where necessary due to breaks. Even then the length of the cable should not be altered as this will change the characteristics of the cable and the heat. It will be necessary to occasionally splice a break, but only approved methods shall be used.

179

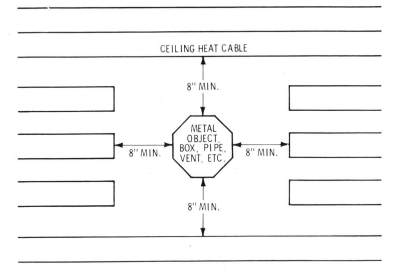

Fig. 112. Clear metal heating cable from metal objects by a minimum of 8 inches.

424-41. Installation of Heating Cables on Dry Board, in Plaster and on concrete ceilings—

(a) **Shall Not Be Installed in Walls.** Heating cable shall not be installed in walls. It is not designed for this purpose and is strictly forbidden, with the exception that isolated runs of cable may be run down a vertical surface to reach a drop ceiling.

(b) **Adjacent Runs.** Adjacent runs of heating cable shall be spaced not closer than 1½ inches on centers and have wattage not to exceed 2¾ watts per square foot. See Fig. 113.

Fig. 113. Heating cable installed in ceilings shall be spaced at least 1½ inches apart.

1-1/2" MIN.

2-3/4 WATTS PER FOOT MAXIMUM

180

(c) **Surfaces to Be Applied.** Heating cables shall be applied only to gypsum board, plaster lath, or similar fire-resistant materials. If applied to the surface of metal lath or any other conducting material, there shall be a coat of plaster commonly known as a brown or scratch coat applied before the cable is installed. This coating of plaster shall entirely cover the metal lath or conducting surface. See **Section 424-41 (f)**.

(d) **Splices.** *All the heating cables, the splice between the heating cable and nonheating leads, and 3 inch minimum of nonheating lead at the splice shall be embedded in plaster or dry board in the same manner as the heating cable.* See Fig. 111.

(e) **Ceiling Surfaces.** On plastered ceilings, the entire surface shall have a finish coat of thermally noninsulating sand plaster or other approved coating which shall have a

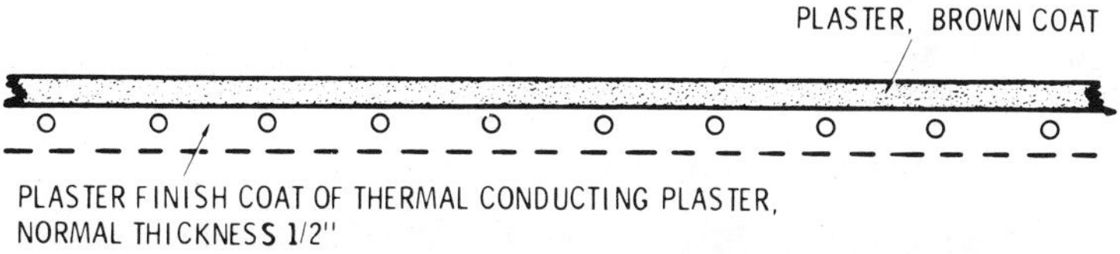

PLASTER, BROWN COAT

PLASTER FINISH COAT OF THERMAL CONDUCTING PLASTER,
NORMAL THICKNESS 1/2"

Fig. 114. Heating cable installed in plaster ceiling.

nominal thickness of ½ inch. Insulation (thermal) plaster shall not be used. See Fig. 114.

(f) **Secured.** The cable shall be fastened at intervals not to exceed 16 inches by means of taping, stapling, or plaster. Staples or metal fasteners which straddle the cable shall not be used with metal lath or other conducting surfaces. The fastening devices shall be of an approved type.

(g) **Dry Board Installations.** When dry board ceilings are used, the cable shall be installed and the entire ceiling below the cable shall be covered with gypsum board not exceeding ½ inch in thickness, but voids between the two layers and around the cable shall be filled with a conducting plaster or other approved thermal conducting material so that the heat will readily transfer. See Fig. 115.

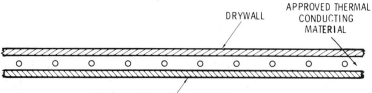

Fig. 115. Heating cable installed in drywall ceiling.

(h) **Free from Contact with Conductive Surfaces.** Heating cables shall not come in contact with metal or other conducting materials.

(i) **Joists.** *In dry board applications, cable shall be installed parallel to the joist, leaving a clear space centered under the joist of 2½ inches (width) between centers of adjacent runs of cables. Crossing of joist by cable shall be kept at a minimum and should be at the ends of the room. Surface layer of gypsum board shall be mounted so that nails or other fastenings do not pierce the heating cable.*

424-42. Finished Ceilings—The question often arises as to whether wallpaper or paint can be used over a ceiling that has heating cable. These materials have been used as finishes over heating cable since cables were first used in the late 1940's. This Section gives formal recognition to painting or papering ceilings.

424-43. Installation of Nonheating Leads of Cables and Panels—

(a) **Free Nonheating Leads.** Only approved wiring methods shall be used for installing the nonheating leads of cables or panels from junction boxes to the underside of the ceiling. In these installations, single leads in raceways (conductors) or single- or multi-conductors Type UF, Type NMC, Type MI or other approved conductors shall be used. Please note the absence of Type NM.

(b) **Lead in Junction Box.** Where nonheating leads terminate in a junction box, there shall not be less than 6 inches of nonheating leads free within the junction box. Also, the markings of the leads shall be visible in the junction box. This is highly important so that the heating cable can be identified. Fig. 116.

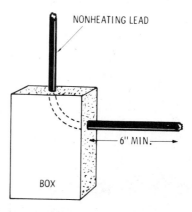

NONHEATING LEAD

6" MIN.

BOX

Fig. 116. Installing nonheating leads in a junction box.

(c) **Excess Leads.** *Excess leads shall not be cut but shall be secured to the underside of the ceiling and embedded in plaster or other approved material, leaving only a length sufficient to reach the junction box with not less than 6 inches of free lead within the box.* See Fig. 116.

(d) **Excess Nonheating Leads.** *Excess nonheating leads of heating panels shall be permitted to be cut to the required length. They shall meet the installation requirements of the wiring methods employed in accordance with* **Section 424-43(a).** *Nonheating leads shall be considered to be an* *integral part of an approved fixed electric space heating panel and not subject to the ampacity requirements of* Section 424-3(b) *for branch circuits.*

424-44. Installation of Panels or Cables in Concrete or Poured Masonry Floors—This Section is for fixed indoor space heating and is not to be confused with ice and snow melting.

(a) **Watts per Square Foot or Linear Foot.** Panels or heating units shall not exceed 33 watts per square foot of heated area or 16½ watts per linear foot of cable.

(b) **Spacing Between Adjacent Runs.** The spacing between adjacent runs of cable shall not be less than 1 inch on centers.

(c) **Secured in Place.** Cables have to be secured in place while concrete or other finish material is being applied. Approved means, such as nonmetallic spreaders of frames, shall be used. Concrete floors often have expansion joints in them. Cables, units, and panels shall be so installed that they do not bridge an expansion joint unless they are protected so as to prevent damage to the cables, units, or panels due to expansion or contraction of the floor.

(d) **Spacing Between Heating Cable and Metal Embedded in the Floor.** Spacings shall be maintained between the heating cable and metal embedded in the floor.
Exception: Grounded metal-clad cable may be in contact with metal embedded in the floor. This included MI heating cable which is being extensively used.

(e) **Leads Protected.** Sleeving of the leads by means of rigid metal conduit, intermediate metal conduit, rigid nonmetallic conduit, electrical metallic tubing, or other approved means shall be used for protection where the leads leave the floor. See Fig. 117.

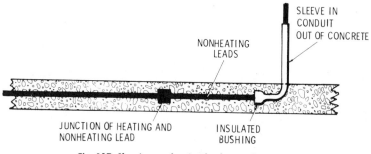

Fig. 117. Sleeving nonheating leads out of concrete.

(f) The sleeves mentioned in (e) shall have bushings or other approved means used where the leads enter or emerge from the floor slab to prevent damage to the cable.

Wiring above heated ceilings shall be located not less than 2 inches above the ceiling and if it is in thermal insulation at this height, it shall be considered as being operating at 50°C and no special protection is required. See Fig. 118.

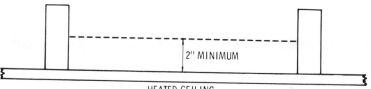

Fig. 118. Conductors shall not be less than 2 inches above heated ceiling.

Personally I would prefer to go above the insulation with the wiring, or at least to the top of the 2 × 4 or 2 × 6 ceiling joists. When ready to install heat cable, draw the room out to scale on paper, get the footage of the cable to be installed and lay it out on the drawing so you can evenly space the cable to use all of it. Then lay out the drawing on the ceiling with a chalk-line.

185

CHAPTER 17

Summary

A summary of what has been covered in the rest of this book might prove to be helpful. In Chapter 2, the BASIS FOR LOAD CALCULATIONS was covered and in Chapter 3, the ACTUAL CALCULATIONS FOR DWELLINGS was covered. Fig. 119 shows a typical layout for wiring a home which is shown as Fig. 6A in Chapter 2. These drawings will be used in this summary as well as the load calculations which were covered in Chapter 2 of this book. Not only will the minimum Code requirements be used, but there will also be suggestions for a more adequate wiring job and additions in the future for additional usage.

The first important consideration is where to locate the service entrance equipment. There are several items which will enter into making a decision:

1. The NEC in **Section 230-72(c)**, states:

(c) **Location.** The disconnecting means shall be located at a readily accessible point nearest the entrance of the conductors, either inside or outside the building or structure. Sufficient access and working space shall be provided about the disconnecting means.

This is short but very meaningful. What is the closest point of entrance? Many inspection authorities consider 15 ft. as a maximum. This, however, varies. *Why the problem?* The service entrance conductors are without over-

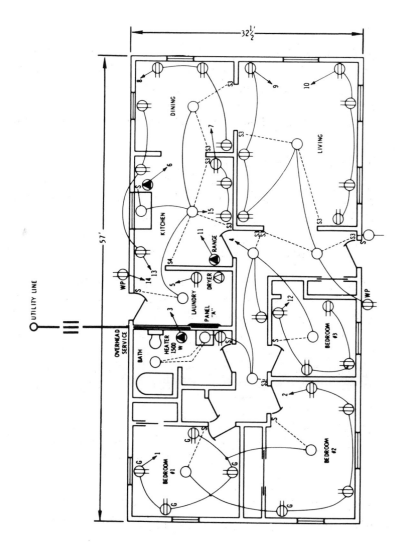

Fig. 119. Service entrance location.

current protection until they arrive at the main disconnect, thus, the length of unprotected conductors must be kept to a minimum for fire and safety protection.

2. A utility must supply service to the residence, therefore they must be considered as to where power lines are located, etc. Thus, one must consult with the utility serving, as to location. If supplied by a service lateral (underground service), the problem eases to quite a degree. The service equipment is at a very good point, if the utility can serve from the point as shown in Fig. 119.

Take a look at Fig. 120. Here the utility line is at the end of the home, which leaves us with several alternatives:

1. The main disconnect and overcurrent device may be located at the end of the house, on the outside in a raintight enclosure (RO) and a feeder circuit (3-wire with equipment grounding conductor) run to the branch circuit panel "A".

2. A service lateral may be run as shown by the dotted or dashed lines in Fig. 120.

These, of course, are not all of the answers, as each case must stand on its own merits, taking into consideration all the points that shall be met.

a. Closest point of entrance, either inside or outside the building.

b. Readily accessible. Do not place the service equipment over a washer or dryer, etc., near combustibles, such as in a closet, in a bathroom or bedroom, or too high.

c. The disconnecting means is for easy access in the case of emergencies.

One thing should be uppermost in the mind of anyone wiring a home, and that is, most people spend the greater

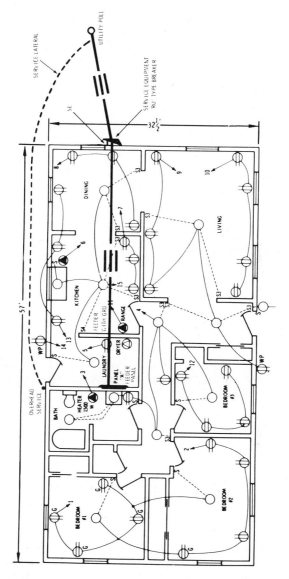

Fig. 120. Two methods of installing service when the utility pole location does not fit equipment location.

189

part of their working life making payments on a home so safety and adequate wiring should be in the mind of every electrician when planning the wiring and in the workmanship used.

What size shall the service be? In Calculations No. 2 and No. 3 (in Chapter 3) we stated the service should be 100 ampere. Remember that this is the minimum requirement. The author, from experience, recommends larger than 100 ampere service (minimum). With the added useages of electricity, we will no doubt overload the 100 ampere service in the future. Plan now to increase the size of the service to a 150 ampere or even in some cases to a 200 ampere service. The added cost is nominal but the cost of increasing the size of the service later will be materially greater. To be competitive on bids, there is nothing to keep you from making a bid on the 100-ampere service and then make an alternate bid for 150 or 200 ampere service and then do a little selling.

Going back to Calculation No. 1, we found that 6- 15 ampere circuits would be required (minimum) for the general lighting load plus a minimum of 2- 20 ampere circuits for the small appliance load and 1- 20 ampere circuit for the laundry circuit; a 35 ampere (two pole) circuit for the range and a 40 ampere (two pole) circuit for the dryer.

6- 15 ampere circuits
3- 20 ampere circuits
2- 35 ampere circuits (figuring 2 poles)
2- 40 ampere circuits (figuring 2 poles)

———

13 circuits minimum

There should also be a few spares for future additions. Standard branch-circuit panels come in the following sizes:

Number of Single poles
12
16
24
30
40
42

The minimum size to be used would be a 16 pole panel. When purchasing the panel, you need not purchase it with breakers to fill all spaces. Look to the future since the base price of the enclosure is a very small part of the total wiring cost, but an addition later will be very costly. Boost your sights and purchase a larger enclosure and thus plan for the future. The installation of the panel readily fit between 16 inches on center studs—secure it solidly and bring it out flush with the finished wall surface. It is easy to install all of the cables or raceways during the rough-in since the walls are open. After the wall is finished, it will be hard to fish in additional circuits. Install a 1¼ inch or 1½ inch conduit or EMT into the attic area and also into the basement area, to facilitate installing any additional circuits that will most certainly be added in the future. See Fig. 121.

It is sometimes impossible to install the conduit into the basement area. In this case, drill the future holes through the plate, etc., and install lengths of single insulated conductors to be used as fish wires for future circuits.

It is amazing how often the inspector hears of a new home owner that has had his house wired only to the minimum Code requirements. The new home owner tells him of additional circuits that he wishes to add but finds it practically impossible to make these additions and meet Code requirements.

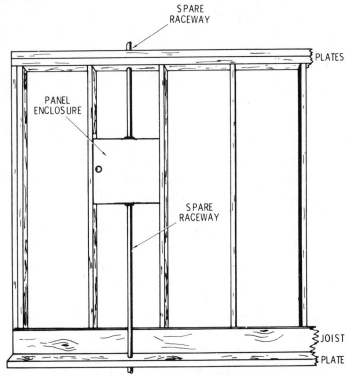

Fig. 121. Spare raceways for future branch circuits.

Bonding and grounding of services was previously covered rather thoroughly in Chapter 5. These have the highest priority. Grounding bars for equipment grounding conductors are available for installation in panels or the enclosures may be purchased with grounding bars installed. See Fig. 122. Review Chapter 5 on SERVICES for additional information on service installation and grounding.

The difference between *feeder panels* and *service entrance panels* with branch-circuit overcurrent devices in the service entrance equipment should be covered. A feeder is the circuit conductors between the service equipment;

or the generator switchboard or an isolated plant, and the branch-circuit overcurrent device.

Most dwellings have the branch-circuit overcurrent devices in the service equipment enclosure as would be the case in Fig. 119. Then in Fig. 120, we have the service equipment outdoors at the end of the house, so the conductors to the feeder panel will be feeders and consist of 2- phase conductors; 1- neutral conductor and 1- green or bare equipment grounding conductor, sized according to **Table 250-95** of the NEC and also appears on page 84 of this book.

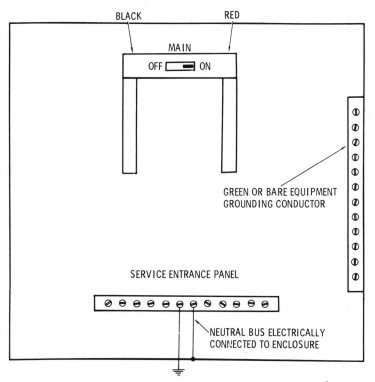

Fig. 122. Equipment Grounding Conductor in a Service Equipment Panel.

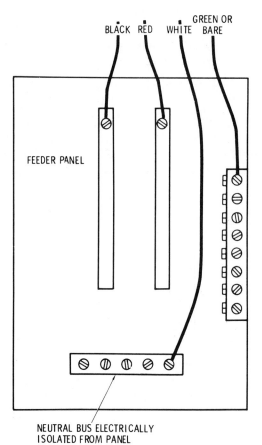

Fig. 123. Feeder panel connections.

Another instance where we could find a feeder is in a large home with the service equipment and branch-circuit breakers in one enclosure, but using a large ampacity breaker to feed a second branch-circuit panel somewhere else about the house.

194

Your attention is called to Fig. 122, which shows the neutral bus electrically connected to the enclosure. This is required in service entrance panels. On feeder panels, we isolate the neutral bus from the enclosure and the green or bare equipment grounding conductor of the feeder (which was just covered) is grounded to the feeder panel enclosure and all branch-circuit equipment grounding conductors are brought to a grounding strap used for this purpose. See Fig. 123.

NUMBER OF OUTLETS PER CIRCUIT

This is a controversial subject, but need not be so. Let us look at what the NEC has to say about it:

The note with **Table 220-2(b)**, tells us about receptacles in single- and multi-family dwellings and further tells us to refer to **Section 220-3(b)**.

Definitions tell us, *A receptacle is a contact device installed at the outlet for connection of a single attachment plug.* It defines a multiple receptacle as *a single device containing two or more receptacles.* From this, when we figure outlets for small appliance circuits, we must consider a duplex receptacle as two outlets and use 180 VA per outlet. A 180 VA may be broken down into amperes: $\frac{180 \text{ VA}}{120 \text{ V}} = 1.5$ amperes per outlet. A small appliance circuit may have 13 outlets or 6 duplex receptacles.

Take a long look at the minimum of two circuits for small appliances and consider adding more than the two circuits even though two looks like enough. In this line of thinking, refer to Fig. 124, you will observe wiring circuits No's. 7 and 13 have been run into the kitchen with No. 13 having one duplex receptacle in the dining room, but there is also circuit No. 8 in the dining room.

195

Looking at Fig. 125, you will notice the room with 4 lights is not indicated as what it is to be used for. It might be a bedroom and if so the receptacle outlets are adequate, but if it is a family room or recreation room then the receptacle in the closet must be put on a lighting circuit, as it would not be a small appliance receptacle.

In Fig. 124, you will notice 2 weatherproof receptacles, one by the front door and one by the rear door of the home. Effective January 1, 1973 these will be required to have *ground-fault circuit-interupters* on these circuits since they are outdoor receptacles. They could both be put on the same circuit. GFIC's are also required for bathroom receptacles.

There has been much progress made in *ground-fault circuit-interupters*. For instance, one company now has a combination interupter and breaker which replaces the ordinary breaker in a panel. There may be other outlets on the interupter besides the outlets that are required by the Code to have an interupter.

In Fig. 125, we see a central air conditioner on circuit No. 19. This may not be installed—the customer may wish to have wall or window air conditioning. If this should be the case, circuit No. 19 could be eliminated and a 120- or 240-volt receptacle circuit run to the location where the wall or window air conditioners will be installed. If the circuit or circuits are for air conditioning only, the circuit may be loaded to 80% of its ampacity, but if other loads, such as lighting are on these circuits the air conditioner may only load the circuit to 50%.

In this summary, nonmetallic sheathed cable will be the principal wiring method shown, as the majority of homes are wired with NM cable. In most instances, wiring with rigid metal conduit or electrical metallic tubing will be very

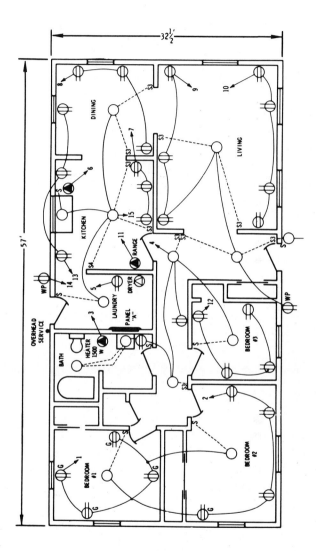

Fig. 124. Circuiting of a house.

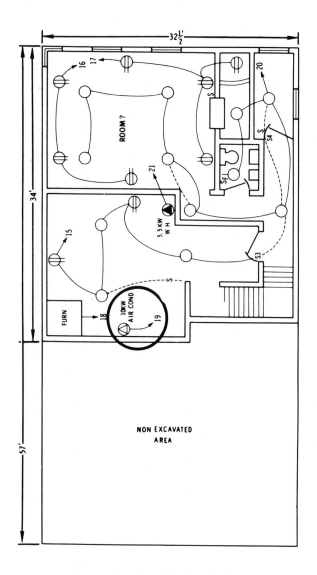

Fig. 125. Lower level circuiting of a house.

198

similar except for the mechanics of installation and that conductors are pulled in after the raceway is installed.

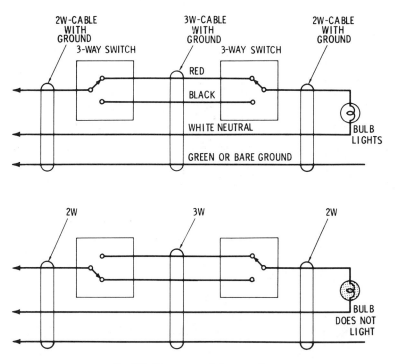

Fig. 126. Three-way switch connections.

Circuits

Three-way and four-way switches seem to cause a little trouble in their connections. Three-way switches are used for turning lights on or off from two places. Three-way and four-way switches are used for turning lights off or on from three or more points. See Figs. 126 and 127.

Referring to Fig. 127, if more than three switching points are required, merely add the extra switches (four-way) in

199

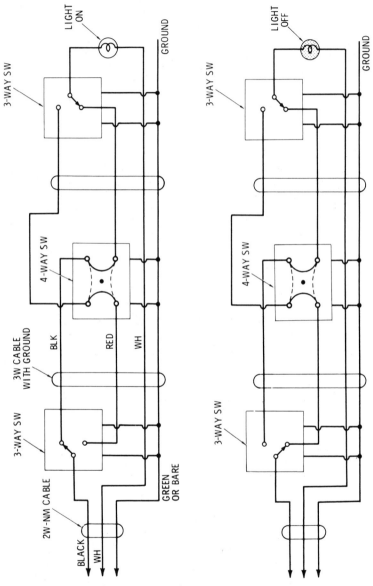

Fig. 127. Two 3-way and one 4-way switch circuit.

the circuit between the two three-way switches. Often the switching of lighting at the garage, is to be done from either the house or the garage. Plus sometimes a hot receptacle may be required in the garage. This connection requires a little more effort, so Fig. 128 is included as a schematic for such a connection.

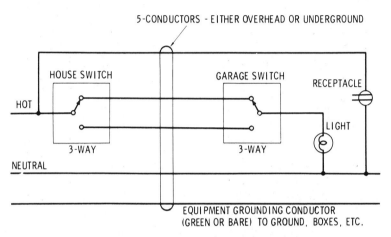

Fig. 128. Three-way switch circuit for garage.

The fifth conductor (equipment grounding conductor) may be eliminated by installing a fuse or breaker-box in the garage for the receptacle(s) circuit. A made electrode is driven in the ground at the garage for the common grounding conductor and connected to both the neutral bus of the overcurrent device in the garage and the breaker enclosure. See Fig. 129.

LOW-VOLTAGE CONTROL WIRING (LV)

Switching of lighting may be done by means of low-voltage controls. This is a very practical means of switching lights on or off. It is simple to install and three-way or four-

201

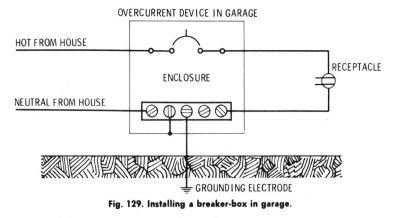

OVERCURRENT DEVICE IN GARAGE

HOT FROM HOUSE

ENCLOSURE

RECEPTACLE

NEUTRAL FROM HOUSE

GROUNDING ELECTRODE

Fig. 129. Installing a breaker-box in garage.

way switches are not required when you wish to switch a circuit from more than one point. One very important purpose in low-voltage wiring is to control most of, or all of, the lights in the home from one common point, such as the master bedroom. Yard lighting may also be controlled by this means. A step-down transformer from 120-volts to 24-volts is designed for this purpose. The transformers are usually designed for about 125 VA capacity and are often mounted in the furnace room or attic.

The conductors usually consist of a 3-conductor low-voltage cable, flat, with one outside conductor ribbed for identification. See Fig. 130. These low-voltage cables need not be installed in conduit, but may be stapled as you would bell wire. Figs. 131, 132, 133 and 134 will illustrate the connections involved. Fig. 131 shows just one light and one switch. Fig. 132 shows two or more switches and one light.

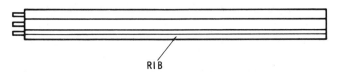

RIB

Fig. 130. Low-voltage switching cable.

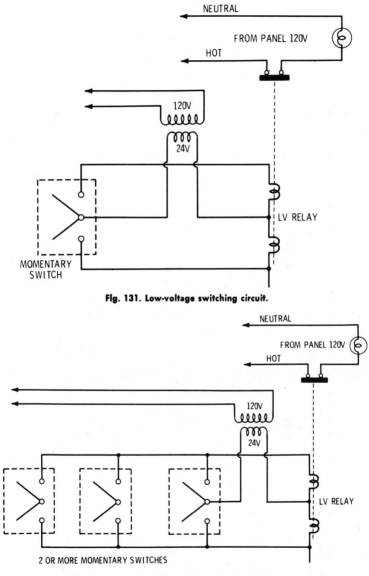

NEUTRAL

FROM PANEL 120V

HOT

120V

24V

MOMENTARY
SWITCH

LV RELAY

Fig. 131. Low-voltage switching circuit.

NEUTRAL

FROM PANEL 120V

HOT

120V

24V

LV RELAY

2 OR MORE MOMENTARY SWITCHES

Fig. 132. Low-voltage circuit showing 2 or more switches for 1 light.

203

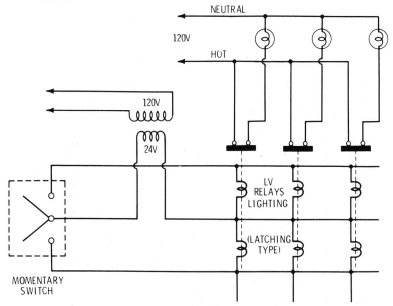

Fig. 133. Low-voltage circuit with 3 or more relays.

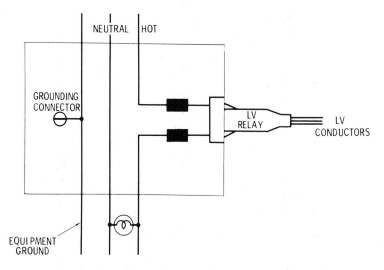

Fig. 134. Typical low-voltage relay installed in a 4-inch square box.

204

Fig. 133 shows one switch and three lights. There are many versions of these connections, but these will illustrate the main points. Fig. 134 shows one of the relays used with low-voltage switching mounted in a junction box. The relay will fit a ½ inch knockout in the box, so the relays may be mounted in individual boxes. A special relay enclosure is made to mount a number of these relays in one enclosure. The primary and secondary sides are always isolated.

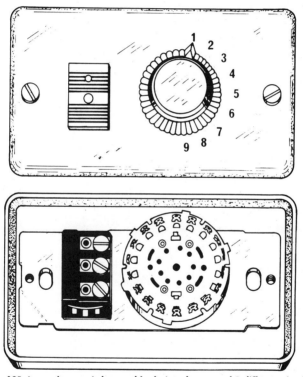

Fig. 135. Low-voltage switch assembly designed to control 9 different circuits.

To control up to nine lights (or circuits) from one location, a 9-point dial-type switch is used. More than one of these switches may be used to accomplish the same thing from more than one point. The connections are parallel and the circuit you wish to control is dialed on the switch and then pushed to the on or off position. See Fig. 135.

CONDUIT

Most homes are wired with NM cable but (even if wired in NM cable) there is usually a certain amount of conduit or EMT which must be installed. The wiring of basement walls for outlets should be laid out and conduit installed

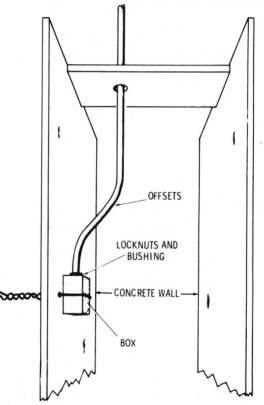

Fig. 136. Conduit and box installation installed before pouring concrete wall.

while the concrete forms are being set for the pouring of the walls. In setting the outlet boxes, do not be concerned with any furring strips used on the walls when they are finished. Box extensions can be added to take care of the furring. When installing conduits for outlets in basement walls, the same outlet spacing will be required as discussed in Chapter 14. The basement may not be finished at the time of occupancy, but it will be later, this you may be assured of.

Fig. 136 illustrates how to rough-in conduit before pouring the concrete. The conduit only needs to be stubbed out above the plate as it will be a raceway for NM cable. Stuff the box with newspaper to assist in cleaning out any concrete that may seep into the box. Conduit may also be run horizontally between outlets, if you so desire.

At most outlets in concealed work, where a house is wired with conduit, a right angle bend is necessary. The inexperienced man has difficulty in making this bend to proper length. In general, the conduit to be bent must have a total depth (Fig. 137A) from the back of the pipe to the end of the bend. The procedure is as follows:

Secure a piece of conduit, and by the aid of a hickey, bend the end up slightly from the floor keeping your left foot on the conduit and close to the hickey. Exert your bending force in two directions; one toward the bend along the line of the hickey handle and the other toward the right foot. When the bend is about three-fourths completed, measure up from the floor to the end of the conduit to determine whether the bend is going to be too short or too long when completed. Shortening or lengthening the bend at this time, if it is necessary, may be done by sliding your hickey up or down on the conduit and continue to bend, being sure to apply the various forces as directed (Fig. 137B).

207

Summary

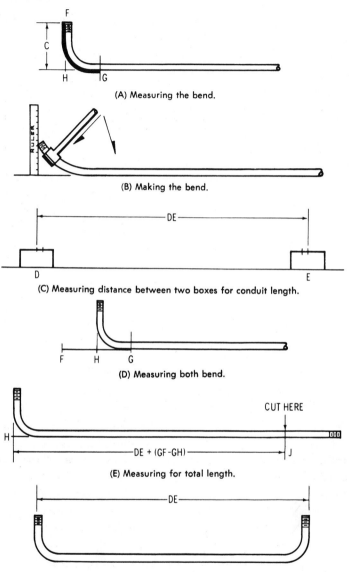

(A) Measuring the bend.

(B) Making the bend.

(C) Measuring distance between two boxes for conduit length.

(D) Measuring both bend.

(E) Measuring for total length.

(F) Measuring total length.

Fig. 137. Method of bending and measuring.

If two outlet boxes are to be connected together by means of a single piece of conduit, proceed as follows: Measure the distance from the far side of the knock-out in box D to the far side of the knock-out box E, Fig. 137C. From the bend (Fig. 137A), determine its length by placing a piece of wire around the bend from F to some point G on the straight part of the pipe. Measure the distance from G to H (Fig. 137A). Subtract the length GH from the total length of the wire GF (Fig. 137D) and add the length HF to the length DE in Fig. 137C. This will give the length of pipe from the back of the first bend to the end of the pipe (Fig. 137E). To illustrate Fig. 137 :

Assume C to B = 8 inches
DE to BE = 6 inches
GH to BE = 5 inches
GF to BE = 12 inches
DE + (GF − GH) = HJ
HJ = (6 inches × 12 inches) + (12 inches − 5 inches)
HJ = 72 inches + 7 inches = 79 inches total.

Care should be exercised in bending the second bend so as to produce a duplicate of the first one. Place the bent end against the wall and measure out on the floor the distance of 6 ft. and bend the conduit so that when finished it will be 6 feet from end to end outside both bends, as shown in Fig. 137F.

Great care should be taken in order to get all conduits and boxes lined up so that when the mechanical work on the home is completed, the electrical devices installed will present a neat appearance. Before installing conductors in the conduit, the conduit should be swabbed out if it is found to contain water. Remember, no conductors are to be installed in conduit until all conduit is complete, and in

209

place, and the home is protected from the elements by the roof, walls, etc.

In a frame home conduit, as a rule, is not used to any great extent. When conduit is installed and concealed, notches are cut in the upper side of the floor just large enough to receive the conduit. *Do not* weaken the floor joists. Care must be taken not to recess the joists, etc., beyond the amount really necessary because of the weakening effect.

A conduit installation is more expensive than a NM cable installation. However, it has the advantage of being more flexible, since conductors may be removed and new conductors substituted or added at any time, provided that the conduit fill has not already been exceeded and any derating required by **Note 8** with **Tables 310-16** through **310-19** are observed.

Refer to Chapter 10 covering ELECTRICAL METALLIC TUBING and Chapter 11 covering RIGID METAL CONDUIT and abide by what is covered in these two chapters. Pay particular attention to Fig. 67, regarding "kinks and wrinkles". Fig. 66 "radius of bends" and Figs. 64 and 65, pertain to the number of bends between pull boxes. Unless you have experienced trouble with pulling conductors or stripping insulation when pulling conductors, this may not mean too much to you. If the rules are not adhered to, such as less bends, you will regret your new experience and it will be rather late to rectify your mistakes, besides you always have an inspector watching for Code violations. Always do a good job of reaming conduit and EMT ends, you fool no one but yourself by any cheating on this rule.

LAYING OUT OF CIRCUITS

You have, or should have, a print of the house and the layout of circuits and outlets on the print. You should also ascertain at what height your customer wishes the outlets and switches to be installed. Check to be sure that the layout of outlets and lighting meets the approval of the home owner. Stay within spacing required by the NEC, but add more outlets if the customer so desires. So often a customer says: "But I do not need an outlet there". Fine, put the outlet where the customer wants it, but if the proper spacing as required by the NEC is not met, explain to the customer the facts in the case. This has happened many times to the author and later the customer has thanked me for the extra outlets that has to be installed to meet Code requirements.

In discussing the heights of receptacles and switches with your customer, the following heights seem to be the most used and may be of assistance to you:

Receptacles 12 inches to 18 inches,
Receptacles over counter-tops 42 inches,
Receptacles in bathrooms over or by sink 42 inches,
Switches 42 inches to 54 inches.

Thermostats should, to be practical, be installed at shoulder height. When installing, it should be at the height most comfortable to attain a temperature most comfortable to your body. You may compensate the setting to attain a comfortable temperature but it seems so much easier to explain the temperature of thermostat setting, if they are as they should be and not to add or subtract degrees in the reaching of a proper temperature at which the customer is comfortable.

After you have the height information for outlets and switches, use a 1 inch × 2 inch board 5 ft. long and mark it off for use in measuring the height of outlets and switches. See Fig. 138. This template, or marking stick, is then used to mark the location of the outlet box. After the out-

211

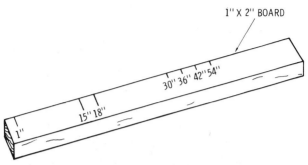

Fig. 138. A template used to mark outlets and switch heights.

lets have been spotted for location, use this board and mark the heights on the studs. Nothing looks worse than "X" number of receptacles in a room at "X" number of heights.

Now that you have locations and heights established, *"What direction will you run the cable to feed these outlets?"* It will be impossible to give you a hard and fast rule for this, but suggestions should be made. I have gone into houses (on inspections) that looked like a spider web. This type of wiring will work but more cable is required and the voltage drop will be greater. The electricity loss plus the extra footage of cable costs money—not only first cost but every day cost. Analyze the wiring layout, a little time spent in doing this will pay large dividends.

The wiring in a house is like "hand writing" to an inspector, he can, in most cases, tell you who wired the house. Pride should be taken in your workmanship, as well as Code requirements.

I call to mind walking into a home to make a rough-in inspection (all walls open), my first impression was "inadequately wired." Then I started checking and found one of the best wired homes that I have ever seen. My first impression was due to the lack of cables run. Time was taken in planning the layout and cable run in so far as practical in a horizontal pattern from outlet to outlet, instead of taking each run to a junction box in the attic or crawl space and thus spider-webbing. I assure you that the most workmanlike wiring of a home that I had ever seen was accomplished with a very minimum footage of cable—you may be assured that this wireman was commended on his work.

Junction boxes are necessary items. Some must, by necessity, be used in attics and crawl spaces, but any extra connections take time and are potential sources of trouble, so the fewer the connections, the better. One thing to remember is that trouble has a way of developing, without any extra assistance, this we must not overlook.

Go back and take a look at Fig. 6 on page 18. The receptacles in bedroom No. 1 were put on a circuit with the light in bedroom No. 2, and by the same token, the receptacles in bedroom No. 2 were put on the same circuit with the light in bedroom No. 1. *Why?* This would leave either the light or receptacles in either bedroom operable in case of trouble on either circuit No. 1 or No. 2. The advantage of this need not be explained, it explains itself, should sickness occur, there is a hot circuit available.

You will also notice that the furnace is on circuit No. 18, in Fig. 6B on page 19. It is not a Code requirement that the furnace be on a separate circuit, but let us suppose that the furnace circuit had other outlets on it and trouble developed, the furnace will not operate and we are without heat. One thing that I did not mention was that a "Fusestat", at the furnace sized to the value required by the furnace motor, should be installed. This is a watchdog for furnace motor trouble, which will tell us that trouble has developed in the motor and we may get the necessary repairs made before a new motor has to be purchased.

As to the dishwasher and disposal, they shall not be put on small appliance circuits. I would suggest a multiwire circuit which is shown in Fig. 139. The definition of a multiwire circuit is: *Branch Circuit, Multiwire: A multiwire branch circuit is a circuit consisting of two or more ungrounded conductors having a potential difference between them, and an identified grounded conductor having equal potential difference between it and each ungrounded conductor of the circuit and which is connected to the neutral conductor of the system.*

The dishwasher is on 1- 120 volt circuit, direct connected and well grounded. Exceptions for this are portable dishwashers which will be plugged into the small appliance circuits. The disposal is on the other 120 volt circuit and well grounded.

One item to remember, is that a disposal is often removed for repairs by someone other than an electrician. If directly connected to the circuit, someone not familiar with equipment grounding conductors, could very easily not appreciate the value of the equipment grounding conductor and miss the proper connection. Personally, I like to see the disposal connected by cord and grounding type attachment plug for easy removal and replacement. This method, of course, may not appeal to you, if not, direct connect it. See Fig. 140.

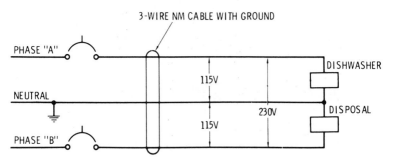

Fig. 139. Multiwire circuit for dishwasher and disposal.

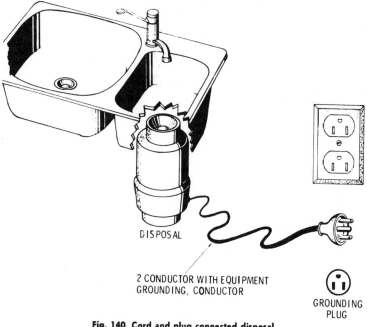

DISPOSAL

2 CONDUCTOR WITH EQUIPMENT
GROUNDING, CONDUCTOR

GROUNDING
PLUG

Fig. 140. Cord and plug connected disposal.

COLOR CODING

Section **200-7** of the NEC tells us that white or natural gray colored conductors are to be used for identified (neutral) conductors and shall be used for no other purpose. Section **200-6** of the NEC tells us: Insulated conductors larger than No. 6, shall have an outer identification of

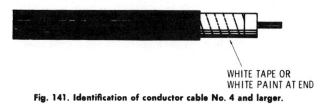

WHITE TAPE OR
WHITE PAINT AT END

Fig. 141. Identification of conductor cable No. 4 and larger.

215

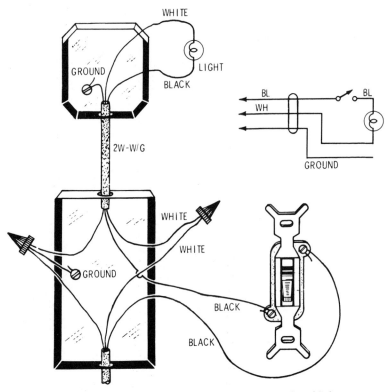

Fig. 142. Color coding for circuit with single-pole switch and light.

white or natural gray color, or shall be identified by distinctive white marking at terminals during process of installation. The reason for permitting marking of conductors larger than No. 6, is that white or natural gray insulation is generally not available in No. 4 and larger conductors. See Fig. 141.

There are exceptions for NM sheathed cable and AC cable because 2-wire cable has 1 black and 1 white, and 3-wire cable has 1 black, 1 red and 1 white conductor. Illustrations for cable markings are shown in Figs. 142 through 146.

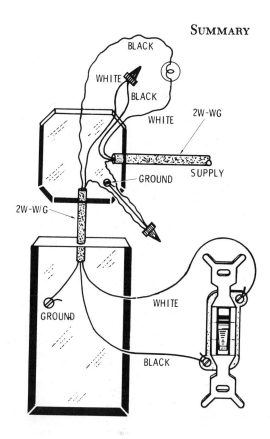

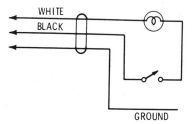

Fig. 143. Color coding for circuit with light at feed and a single-pole switch.

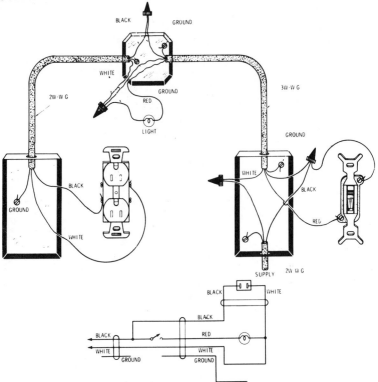

Fig. 144. Coding for switched light and hot receptacle.

From these illustrations, you may readily see that white (neutral) goes to the light or the screw connection on screw-base sockets. Elsewhere, the white may be used as a traveler but then a black wire shall go back to the light. A neutral (identified conductor) is never connected to a switch. There is nothing to prohibit you from color-coding off colors at switching points, in fact it is recommended by the author, should you so desire.

BOX FILL

It might be well to review box fill, using the illustrations shown in Figs. 142 through 146. In Fig. 142, we have two

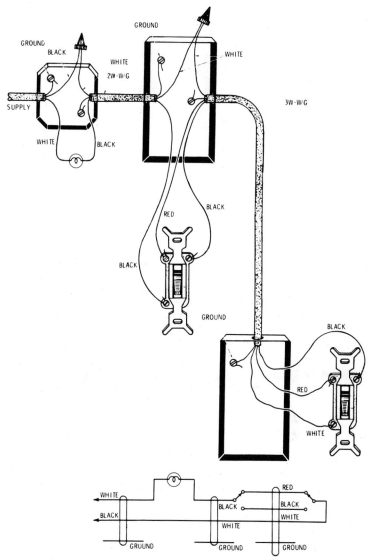

Fig. 145. Coding for feed to light with two 3-way switches.

Summary

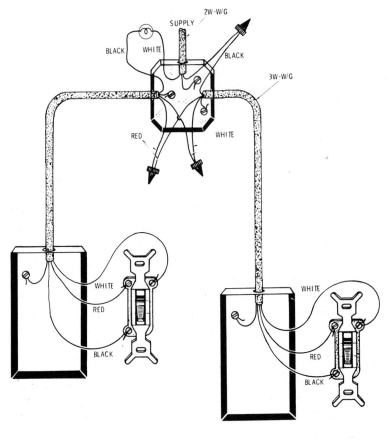

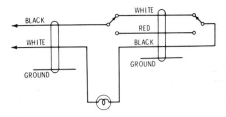

Fig. 146. Coding for feed to light with two 3-way switches.

2-wire cables with ground connected to the switch box. Thus, we have a total of 4 current carrying conductors entering the box, plus 2 equipment grounding conductors, making a total of 6 conductors. Then we have cable clamps and one device (switch). The grounding conductors (2) are counted as one conductor, plus 4 current carrying conductors or 5 conductors. We add the cable clamps and one for the switch, making 7 conductors total, for figuring fill. If these are No. 14's, using **Table 370-6(a)** on page 154, we discover it will take a 3 × 2 × 2¾ device box. If No. 12's were used, we find that it would take a 4 × 1½ box.

In the octagon box, for the light, we have 2-wire cable with ground or 3 conductors and cable clamps. There would be no device, but if a fixture stud was used for mounting the light fixture, we would have three conductors, cable clamps and a fixture stud, making a total of 5 conductors to use in calcuating the box size. With No. 14's, using **Table 370-6(a)**, we could use either a 3¼ × 1½ octagonal box, a 3½ × 1½ octagonal box or a 4 × 1½ octagonal box. If No. 12's were used we could use the 3½ × 1½ octagonal box or a 4 × 1½ octagonal box.

This does not mean that you must use the exact box, you may use larger boxes but never smaller boxes. Since boxes for wiring a home are purchased in quantity, you may prefer the larger box to simplify the number of different size boxes that you use.

Take the octagon box in Fig. 146. Here we have two 3-wire cables and one 2- wire cable, with grounds. Thus, we have 8 current carrying conductors and three grounding conductors, but we are only required to count 1 grounding conductor, so we have 9 conductors to figure—plus cable clamps and fixture stud, if used, or 11 conductors.

If No. 14's are used, from **Table 370-6(a)**, we find we need a 4 × 2⅛ octagonal box. If No. 12's were used, we would need a 4 × 2⅛ square box or a 4 × 2⅛ octagonal box with an extension, as covered in Fig. 92, Page 151.

INSTALLATION OF CABLES

Allowance must be made to allow cable movement as a house settles. In Figs. 43, 44, and 45 (page 89 and 90), you will notice that the staples were not placed right at the cable bends, this allows for freedom of movement. Also, never install the staple too tight, this is because you might cut the insulation.

Most homes have an attic scuttle hole for access into the attic. **Section 333-12** covering AC Cable will be copied here as it also applies to NM sheathed cable.

333-12. In Accessible Attics—*Type AC cables in accessible attics or roof spaces shall be installed as follows:*

(a) **Where Run Across the Top of Floor Joists.** *Where run across the top of floor joist, or within 7 feet of floor or floor joists across the face of rafters or studding, in attics and roof spaces which are accessible, the cable shall be protected by substantial guard strips which are at least*

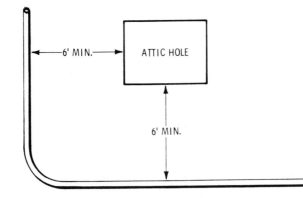

Fig. 147. Cable in attic not protected by running boards.

as high as the cable. *Where this space is not accessible by permanent stairs or ladders, protection will only be required within 6 feet of the nearest edge scuttle hole or attic entrance.*

(b) **Where Carried Along the Sides of Floor Joists.** *Where cable is carried along the sides of rafters, studs or floor joists, neither guard strips nor running boards shall be required.* See Figs. 147 and 148.

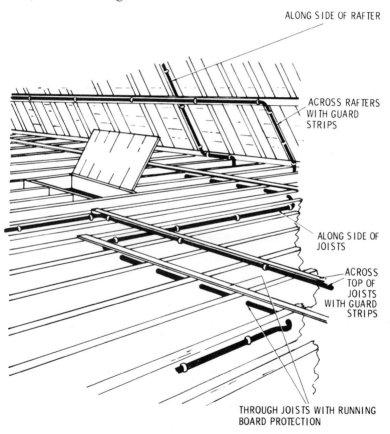

Fig. 148. Method of running nonmetallic sheathed cable in attic.

From these illustrations, you see that there are two alternatives: (1) keep the cable a minimum of 6 feet from the scuttle hole, and (2) if run is closer than 6 ft., install running boards for protection of the cable.

BOX AND DEVICE GROUNDING

We should never lose sight of the importance of the equipment grounding conductors. Therefore we must take every precaution to see that the equipment grounding conductors are electrically continuous and properly made, with

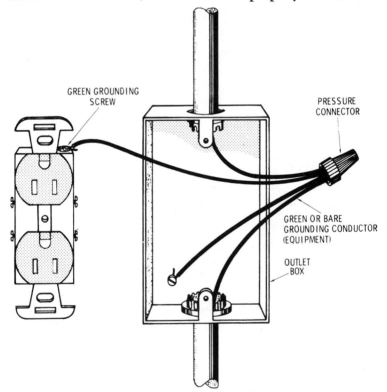

GREEN GROUNDING SCREW

PRESSURE CONNECTOR

GREEN OR BARE GROUNDING CONDUCTOR (EQUIPMENT)

OUTLET BOX

Fig. 149. Proper connections for equipment grounding conductors.

low impedance, so that they may serve the purpose for which they were intended.

What happens if a phase conductor shorts to ground in a box, or an electric drill, hedge clipper, saw, etc., develops a ground? The box appliance enclosure develops a potential above ground, which will pass through your body. The equipment grounding conductors are paths for this potential (above ground) to return to the fuse or breaker protecting the circuit involved, blowing the fuse or tripping the breaker, thus, removing this potential above ground.

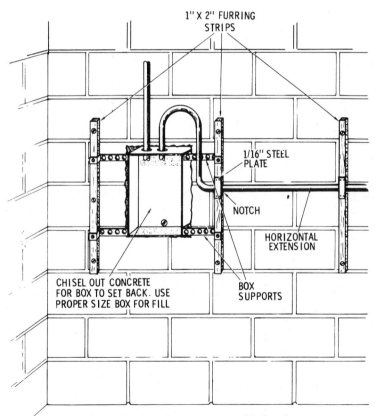

Fig. 150. Furring strips on concrete or block walls.

SUMMARY

In making up grounds in boxes where you have more than one grounding conductor, do not place more than one grounding conductor under a box grounding screw. Splice all conductors together by means of an approved connector, installing a pigtail for the box grounding and another pigtail for the connection to the green grounding screw of the switch, receptacle, or other device. See Fig. 149. All boxes have tapped holes for $10\frac{}{32}$ grounding screws. Purchase the approved large headed grounding screws, never use solder.

There are "UL" approved devices such as receptacles that ground to the box and a grounding conductor will not be required. There are also "UL" approved grounding clips used for grounding the equipment grounding conductor to the box edge.

WIRING BASEMENTS

Earlier in this Chapter, the rough-in boxes in concrete basement walls were discussed. This of course covered houses under construction. Sometime you may be faced with installing outlets in the finished basement (concrete) wall.

If conduit was installed in the concrete walls during construction, you will find the job of wiring simple. If there is no conduit, the wiring must be installed flush with the wall. Many people like to furr out the basement walls and install drywall or paneling. This will simplify the installation of the necessary outlets to meet the NEC requirements.

Furring is usually accomplished by nailing 1- or 1½-inch strips to the concrete or block walls and the drywall or paneling installed on these strips. See Fig. 150. Chisel out enough concrete to accommodate a device box of the proper cubic inch capacity and support the box with box hangers. This will be required due to the damp location.

You may also continue with horizontal runs by cutting out notches in furring strips for the cable. Cover the cutout portion with ⅟₁₆ inch steel plate, to protect the cable from nails.

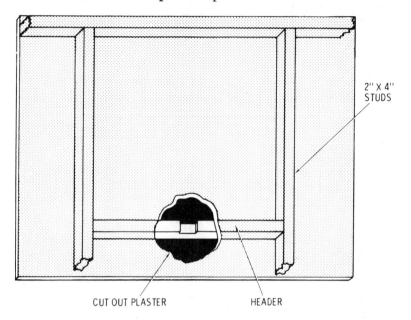

Fig. 151. Notching out old headers (fire stops) for new electrical wiring.

ADDING WIRE TO OLD HOMES

Here is where you use your ingenuity. No book would be large enough to possibly cover all possible problems which you might encounter. Some of the more common problems will be covered.

If the older house has lathe and plaster, the old plaster is often removed and replastered or drywall is installed over the old plaster. In these cases your problem will not be large, as you may cut holes in the walls and fish in the NM cable very readily.

SUMMARY

A very common problem that one runs into is headers (fire stops) between joists. This is usually overcome as illustrated in Fig. 151. The header is located and plaster chipped out. The header is notched and covered with a $\frac{1}{16}$ inch steel plate after the NM cable has been installed and then patched with plaster. If the wall is brick, cut a channel 2 inches deep in the brick, install NMC cable or UF cable at the bottom of the chase, install a strip of $\frac{1}{16}$ inch steel over the cable and replaster.

Should you wish to fish down partitions between wall studs, it is generally best to use a small metal sash chain because of its flexibility. Locate the point at which you wish to install the outlet or switch, drill the top plates from the attic and drop the chain through this hole. It usually can be heard rubbing the wall. Cut an opening, fish the chain with a bent piece of wire and pull in the NM sheathed cable.

Appendix

(MOT)	Electric Motor	(46.3)
(WH)	Electric Watt-hour Meter	(48)
	Circuit Breaker	(11.1)
	Fusible Element	(36)
	Single-Throw Knife Switch	(76.3)
	Double-Throw Knife Switch	(76.2)
	Ground	(13.1)
	Battery	(7)

APPENDIX

1.0		**LIGHTING OUTLETS**
	Ceiling	**Wall**

1.1 ○ —○ Surface or Pendant Incandescent. Mercury Vapor or Similar Lamp Fixture

1.2 Ⓡ —Ⓡ Recessed Incandescent Mercury Vapor or Similar Lamp Fixture

1.3 ▭○▭ Surface or Pendant Individual Fluorescent Fixture

1.4 ▭○R▭ Recessed Individual Fluorescent Fixture

1.5 ▭○▭▭ Surface or Pendant Continuous-Row Fluorescent Fixture

1.6 ▭○R▭▭ [1]Recessed Continuous-Row Fluorescent Fixture

1.7 ⊢——+——+——⊣ [2]Bare-Lamp Fluorescent Strip

1.10 Ⓑ —Ⓑ Blanket Outlet

1.11 Ⓙ —Ⓙ Junction Box

1.12 Ⓛ —Ⓛ Outlet Controlled by Low-Voltage Switching When Relay Is Installed in Outlet Box

2.0		**RECEPTACLE OUTLETS**
	Ungrounded	**Grounding**

2.1 ⊖ ⊖G Single Receptacle Outlet

2.2 ⊜ ⊜G Duplex Receptacle Outlet

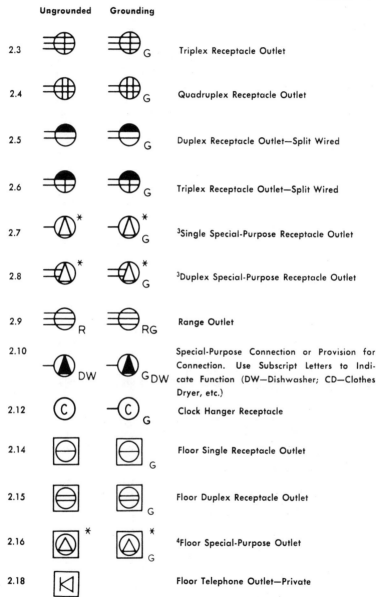

	Ungrounded	Grounding	
2.3		G	Triplex Receptacle Outlet
2.4		G	Quadruplex Receptacle Outlet
2.5		G	Duplex Receptacle Outlet—Split Wired
2.6		G	Triplex Receptacle Outlet—Split Wired
2.7	*	* G	[3]Single Special-Purpose Receptacle Outlet
2.8	*	* G	[3]Duplex Special-Purpose Receptacle Outlet
2.9	R	RG	Range Outlet
2.10	DW	G DW	Special-Purpose Connection or Provision for Connection. Use Subscript Letters to Indicate Function (DW—Dishwasher; CD—Clothes Dryer, etc.)
2.12	C	C G	Clock Hanger Receptacle
2.14		G	Floor Single Receptacle Outlet
2.15		G	Floor Duplex Receptacle Outlet
2.16	*	* G	[4]Floor Special-Purpose Outlet
2.18			Floor Telephone Outlet—Private

231

APPENDIX

3.0		**SWITCH OUTLETS**
3.1	S	Single-Pole Switch
3.2	S_2	Double-Pole Switch
3.3	S_3	Three-Way Switch
3.4	S_4	Four-Way Switch
3.5	S_K	Key-Operated Switch
3.6	S_P	Switch and Pilot Lamp
3.7	S_L	Switch for Low-Voltage Switching System
3.8	S_{LM}	Master Switch for Low-Voltage Switching System
3.9	 —⊖S	Switch and Single Receptacle
3.10	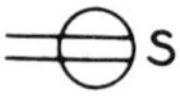 =⊖S	Switch and Double Receptacle
3.11	S_D	Door Switch
3.12	S_T	Time Switch
3.13	S_{CB}	Circuit Breaker Switch
3.14	S_{MC}	Momentary Contact Switch or Push-button for Other Than Signalling System
3.15	Ⓢ	Ceiling Pull Switch

232

4.0

INSTITUTIONAL, COMMERCIAL, AND INDUS-TRIAL OCCUPANCIES

4.9

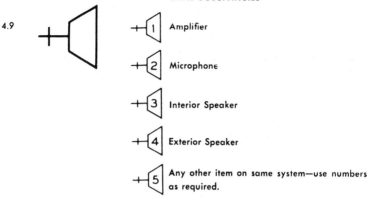

Amplifier

Microphone

Interior Speaker

Exterior Speaker

Any other item on same system—use numbers as required.

5.0

RESIDENTIAL OCCUPANCIES

5.1   Push-button

5.2   Buzzer

5.3   Bell

5.4   Combination Bell-Buzzer

5.5   Chime

5.6   Annunicator

5.7   Electric Door Opener

5.8   Maid's Signal Plug

5.9   Interconnection Box

5.10   Bell-Ringing Transformer

5.11   Outside Telephone

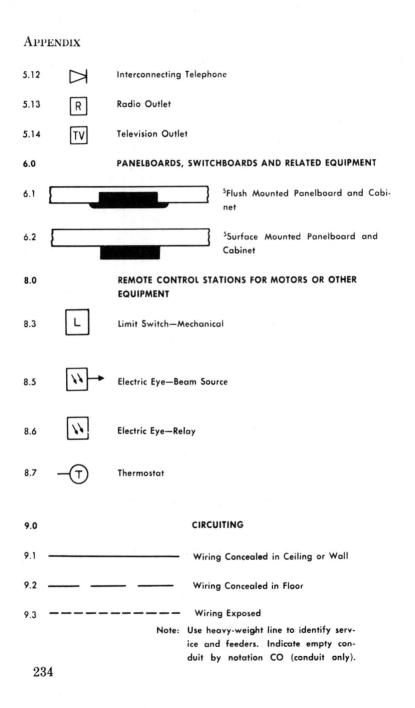

5.12   Interconnecting Telephone

5.13   R   Radio Outlet

5.14   TV   Television Outlet

6.0   PANELBOARDS, SWITCHBOARDS AND RELATED EQUIPMENT

6.1   [5]Flush Mounted Panelboard and Cabinet

6.2   [5]Surface Mounted Panelboard and Cabinet

8.0   REMOTE CONTROL STATIONS FOR MOTORS OR OTHER EQUIPMENT

8.3   L   Limit Switch—Mechanical

8.5   Electric Eye—Beam Source

8.6   Electric Eye—Relay

8.7   T   Thermostat

9.0   CIRCUITING

9.1   Wiring Concealed in Ceiling or Wall

9.2   Wiring Concealed in Floor

9.3   Wiring Exposed

Note: Use heavy-weight line to identify service and feeders. Indicate empty conduit by notation CO (conduit only).

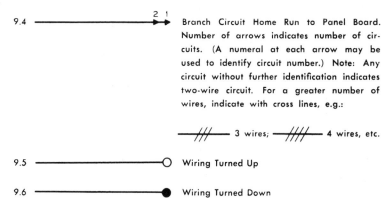

9.4 — Branch Circuit Home Run to Panel Board. Number of arrows indicates number of circuits. (A numeral at each arrow may be used to identify circuit number.) Note: Any circuit without further identification indicates two-wire circuit. For a greater number of wires, indicate with cross lines, e.g.:

——///—— 3 wires; ——////—— 4 wires, etc.

9.5 ——————O Wiring Turned Up

9.6 ——————● Wiring Turned Down

NOTES

[1]In the case of combination continuous-row fluorescent and incandescent spotlights, use combinations of the above standard symbols.

[2]In the case of continuous-row bare-lamp fluorescent strip above an area-wide diffusing means, show each fixture run, using the standard symbol; indicate area of diffusing means and type by light shading and/or drawing notation.

[3]Use numeral or letter either within the symbol or as a subscript alongside the symbol keyed to explanation in the drawing list of symbols to indicate type of receptacle or usage.

[3]Use numeral or letter either with the symbol or as a subscript alongside the symbol

[5]Identify by notation or schedule.

235

Table 8. Properties of Conductors

Size AWG MCM	Area Cir. Mils	Concentric Lay Stranded Conductors			Bare Conductors		D. C. Resistance Ohms/M Ft. At 25°C. 77°F.		
		No. Wires	Diam. Each Wire Inches		Diam. Inches	*Area Sq. Inches	Copper		Alumni-num
							Bare Cond.	Tin'd. Cond.	
18	1620	Solid	.0403		.0403	.0013	6.51	6.79	10.7
16	2580	Solid	.0508		.0508	.0020	4.10	4.26	6.72
14	4110	Solid	.0641		.0641	.0032	2.57	2.68	4.22
12	6530	Solid	.0808		.0808	.0051	1.62	1.68	2.66
10	10380	Solid	.1019		.1019	.0081	1.018	i.06	1.67
8	16510	Solid	.1285		.1285	.0130	.6404	.659	1.05
6	26240	7	.0612		.184	.027	.410	.427	.674
4	41740	7	.0772		.232	.042	.259	.269	.424
3	52620	7	.0867		.260	.053	.205	.213	.336
2	66350	7	.0974		.292	.067	.162	.169	.266
1	83690	19	.0664		.332	.087	.129	.134	.211
0	105600	19	.0745		.372	.109	.102	.106	.168
00	133100	19	.0837		.418	.137	.0811	.0843	.133
000	167800	19	.0940		.470	.173	.0642	.0668	.105
0000	211600	19	.1055		.528	.219	.0509	.0525	.0836
250	250000	37	.0822		.575	.260	.0431	.0449	.0708
300	300000	37	.0900		.630	.312	.0360	.0374	.0590
350	350000	37	.0973		.681	.364	.0308	.0320	.0505
400	400000	37	.1040		.728	.416	.0270	.0278	.0442
500	500000	37	.1162		.813	.519	.0216	.0222	.0354
600	600000	61	.0992		.893	.626	.0180	.0187	.0295
700	700000	61	.1071		.964	.730	.0154	.0159	.0253
750	750000	61	.1109		.998	.782	.0144	.0148	.0236
800	800000	61	.1145		1.030	.833	.0135	.0139	.0221
900	900000	61	.1215		1.090	.933	.0120	.0123	.0197
1000	1000000	61	.1280		1.150	1.039	.0108	.0111	.0177
1250	1250000	91	.1172		1.289	1.305	.00863	.00888	.0142
1500	1500000	91	.1284		1.410	1.561	.00719	.00740	.0118
1750	1750000	127	.1174		1.526	1.829	.00616	.00634	.0101
2000	2000000	127	.1255		1.630	2.087	.00539	.00555	.00885

* Area given is that of a circle having a diameter equal to the overall diameter of a stranded conductor.

The values given in the Table are those given in Handbook 100 of the National Bureau of Standards except that those shown in the 8th column are those given in Specification B33 of the American Society for Testing and Materials, and those shown in the 9th column are those given in Standard No. S-19-81 of the Insulated Power Cable Engineers Association and Standard No. WC3-1964 of the National Electrical Manufacturers Association.

The resistance values given in the last three columns are applicable only to direct current. When conductors larger than No. 4/0 are used with alternating current the multiplying factors in Table 9, Chapter 9 should be used to compensate for skin effect.

Table 5. Dimensions of Rubber-Covered and Thermoplastic-Covered Conductors

Size AWG MCM	Types RF-2, RFH-2, RH, RHH,*** RHW,*** SF-2		Types TF, T, THW,† TW, RUH,† RUW**		Types TFN, THHN, THWN		Types FEP, FEPB, PF, PGF		Type XHHW	
	Approx. Diam. Inches	Approx. Area Sq. In.	Approx. Diam. Inches	Approx. Area Sq. In.	Approx. Diam. Inches	Approx. Area Sq. In.	Approx. Diam. Inches	Approx. Area Sq. Inches	Approx. Diam. Inches	Approx. Area Sq. In.
Col. 1	Col. 2	Col. 3	Col. 4	Col. 5	Col. 6	Col. 7	Col. 8	Col. 9	Col. 10	Col. 11
18	.146	.0167	.106	.0088	.089	.0064	.081	.0052	...	...
16	.158	.0196	.118	.0109	.100	.0079	.092	.0066	...	...
14	2/64 in. .171	.0230	.131	.0135	.105	.0087	.105 .105	.0087 .0087	...	...
14	3/64 in. .204*	.0327*	...	...	...	...	...	...	.129	.0131
14	...	...	.162†	.0206†	...	...	...	...	...	...
12	2/64 in. .188	.0278	.148	.0172	.122	.0117	.121 .121	.0115 .0115	.146	.0167
12	3/64 in. .221*	.0384*	...	...	...	...	...	...	...	...
12	...	...	.179†	.0251†	...	...	...	...	...	...
10	.242	.0460	.168	.0224	.153	.0184	.142 .142	.0159 .0159	.166	.0216
10	...	...	.199†	.0311†	...	...	...	...	...	...
8	.311	.0760	.228	.0408	.201	.0317	.189 .169	.0280 .0225	.224	.0394
8	...	...	.259†	.0526†	...	...	...	...	...	...
6	.397	.1238	.323	.0819	.257	.0519	.244 .302	.0467 .0716	.282	.0625
4	.452	.1605	.372	.1087	.328	.0845	.292 .350	.0669 .0962	.328	.0845
3	.481	.1817	.401	.1263	.356	.0995	.320 .378	.0803 .1122	.356	.0995
2	.513	.2067	.433	.1473	.388	.1182	.352 .410	.0973 .1316	.388	.1182
1	.588	.2715	.508	.2027	.450	.1590	...	...	.450	.1590
0	.629	.3107	.549	.2367	.491	.1893	...	...	.491	.1893
00	.675	.3578	.595	.2781	.537	.2265	...	...	.537	.2265
000	.727	.4151	.647	.3288	.588	.2715	...	...	.588	.2715
0000	.785	.4840	.705	.3904	.646	.3278	...	...	.646	.3278

* The dimensions of Types RHH and RHW.
** No. 14 to No. 2.
† Dimensions of THW in sizes 14 to 8. No. 6 THW and larger is the same dimension as T.
*** Dimensions of RHH and RHW without outer covering are the same as THW.
No. 18 to No. 8, solid; No. 6 and larger, stranded.
**** In Columns 8 and 9 the values shown for sizes No. 1 thru 0000 are for TFE only. The right-hand values in Columns 8 and 9 are for FEPB only.

Table 5-Continued

Size AWG MCM	Types RF-2, RFH-2, RH, RHH,*** RHW,*** SF-2		Types TF, T, THW,† TW, RUH,** RUW**		Types TFN, THHN, THWN		Types **** FEP, FEPB, PF, PGF, TFE, PTF		Type XHHW	
	Approx. Diam. Inches	Approx. Area Sq. In.	Approx. Diam. Inches	Approx. Area Sq. In.	Approx. Diam. Inches	Approx. Area Sq. In.	Approx. Diam. Inches	Approx. Area Sq. Inches	Approx. Diam. Inches	Approx. Area Sq. In.
Col. 1	Col. 2	Col. 3	Col. 4	Col. 5	Col. 6	Col. 7	Col. 8	Col. 9	Col. 10	Col. 11
250	.868	.5917	.788	.4877	.716	.4026	…	…	.716	.4026
300	.933	.6837	.843	.5581	.771	.4669	…	…	.771	.4669
350	.985	.7620	.895	.6291	.822	.5307	…	…	.822	.5307
400	1.032	.8365	.942	.6969	.869	.5931	…	…	.869	.5931
500	1.119	.9834	1.029	.8316	.955	.7163	…	…	.955	.7163
600	1.233	1.1940	1.143	1.0261	1.058	.8792	…	…	1.073	.9043
700	1.304	1.3355	1.214	1.1575	1.129	1.0011	…	…	1.145	1.0297
750	1.339	1.4082	1.249	1.2252	1.163	1.0623	…	…	1.180	1.0936
800	1.372	1.4784	1.282	1.2908	1.196	1.1234	…	…	1.210	1.1499
900	1.435	1.6173	1.345	1.4208	1.259	1.2449	…	…	1.270	1.2668
1000	1.494	1.7531	1.404	1.5482	1.317	1.3623	…	…	1.330	1.3893
1250	1.676	2.2062	1.577	1.9532	…	…	…	…	1.500	1.7672
1500	1.801	2.5475	1.702	2.2748	…	…	…	…	1.620	2.0612
1750	1.916	2.8895	1.817	2.5930	…	…	…	…	1.740	2.3779
2000	2.021	3.2079	1.922	2.9013	…	…	…	…	1.840	2.6590

Table 6. Dimensions of Lead-Covered Conductors
Types RL, RHL, and RUL

Size AWG-MCM	Single Conductor		Two Conductor		Three Conductor	
	Diam. Inches	Area Sq. Ins.	Diam. Inches	Area Sq. Ins.	Diam. Inches	Area Sq. Ins.
14	.28	.062	.28 x .47	.115	.59	.273
12	.29	.066	.31 x .54	.146	.62	.301
10	.35	.096	.35 x .59	.180	.68	.363
8	.41	.132	.41 x .71	.255	.82	.528
6	.49	.188	.49 x .86	.369	.97	.738
4	.55	.237	.54 x .96	.457	1.08	.916
2	.60	.283	.61 x 1.08	.578	1.21	1.146
1	.67	.352	.70 x 1.23	.756	1.38	1.49
0	.71	.396	.74 x 1.32	.859	1.47	1.70
00	.76	.454	.79 x 1.41	.980	1.57	1.94
000	.81	.515	.84 x 1.52	1.123	1.69	2.24
0000	.87	.593	.90 x 1.64	1.302	1.85	2.68
250	.98	.754	. . . .	. . . . .	2.02	3.20
300	1.04	.85	. . . .	. . . . .	2.15	3.62
350	1.10	.95	. . . .	. . . . .	2.26	4.02
400	1.14	1.02	. . . .	. . . . .	2.40	4.52
500	1.23	1.18	. . . .	. . . . .	2.59	5.28

The above cables are limited to straight runs or with nominal offsets equivalent to not more than two quarter bends.

Note — No. 14 to No. 8, solid conductors: No. 6 and larger, stranded conductors. Data for 30-mil insulation not yet compiled.

Table 7. Dimensions of Asbestos-Varnished-Cambric Insulated Conductors

Types AVA, AVB, and AVL

Size AWG MCM	Type AVA		Type AVB		Type AVL	
	Approx. Diam. Inches	Approx. Area Sq. Ins.	Approx. Diam. Inches	Approx. Area Sq. Ins.	Approx. Diam. Inches	Approx. Area Sq. Ins.
14	.245	.047	.205	.033	.320	.080
12	.265	.055	.225	.040	.340	.091
10	.285	.064	.245	.047	.360	.102
8	.310	.075	.270	.057	.390	.119
6	.395	.122	.345	.094	.430	.145
4	.445	.155	.395	.123	.480	.181
2	.505	.200	.460	.166	.570	.255
1	.585	.268	.540	.229	.620	.300
0	.625	.307	.580	.264	.660	.341
00	.670	.353	.625	.307	.705	.390
000	.720	.406	.675	.358	.755	.447
0000	.780	.478	.735	.425	.815	.521
250	.885	.616	.855	.572	.955	.715
300	.940	.692	.910	.649	1.010	.800
350	.995	.778	.965	.731	1.060	.885
400	1.040	.850	1.010	.800	1.105	.960
500	1.125	.995	1.095	.945	1.190	1.118
550	1.165	1.065	1.135	1.01	1.265	1.26
600	1.205	1.140	1.175	1.09	1.305	1.34
650	1.240	1.21	1.210	1.15	1.340	1.41
700	1.275	1.28	1.245	1.22	1.375	1.49
750	1.310	1.35	1.280	1.29	1.410	1.57
800	1.345	1.42	1.315	1.36	1.440	1.63
850	1.375	1.49	1.345	1.43	1.470	1.70
900	1.405	1.55	1.375	1.49	1.505	1.78
950	1.435	1.62	1.405	1.55	1.535	1.85
1000	1.465	1.69	1.435	1.62	1.565	1.93

Note: No. 14 to No. 8, solid, No. 6 and larger, stranded; except AVL where all sizes are stranded.

Table 9. Multiplying Factors for Converting DC Resistance to 60-Hertz AC Resistance

Size	Multiplying Factor			
	For Nonmetallic Sheathed Cables in Air or Nonmetallic Conduit		For Metallic Sheathed Cables or all Cables in Metallic Raceways	
	Copper	Aluminum	Copper	Aluminum
Up to 3 AWG	1.	1.	1.	1.
2	1.	1.	1.01	1.00
1	1.	1.	1.01	1.00
0	1.001	1.000	1.02	1.00
00	1.001	1.001	1.03	1.00
000	1.002	1.001	1.04	1.01
0000	1.004	1.002	1.05	1.01
250 MCM	1.005	1.002	1.06	1.02
300 MCM	1.006	1.003	1.07	1.02
350 MCM	1.009	1.004	1.08	1.03
400 MCM	1.011	1.005	1.10	1.04
500 MCM	1.018	1.007	1.13	1.06
600 MCM	1.025	1.010	1.16	1.08
700 MCM	1.034	1.013	1.19	1.11
750 MCM	1.039	1.015	1.21	1.12
800 MCM	1.044	1.017	1.22	1.14
1000 MCM	1.067	1.026	1.30	1.19
1250 MCM	1.102	1.040	1.41	1.27
1500 MCM	1.142	1.058	1.53	1.36
1750 MCM	1.185	1.079	1.67	1.46
2000 MCM	1.233	1.100	1.82	1.56

241

Decimal Inch Equivalents of
Millimeters and Fractional Parts of Millimeters

mm		Inches	mm		Inches	mm		Inches	mm		Inches
1-100	=	.00039	33-100	=	.01299	64-100	=	.02520	95-100	=	.03740
2-100	=	.00079	34-100	=	.01339	65-100	=	.02559	96-100	=	.03780
3-100	=	.00118	35-100	=	.01378	66-100	=	.02598	97-100	=	.03819
4-100	=	.00157	36-100	=	.01417	67-100	=	.02638	98-100	=	.03858
5-100	=	.00197	37-100	=	.01457	68-100	=	.02677	99-100	=	.03898
6-100	=	.00236	38-100	=	.01496	69-100	=	.02717	1	=	.03937
7-100	=	.00276	39-100	=	.01535	70-100	=	.02756	2	=	.07874
8-100	=	.00315	40-100	=	.01575	71-100	=	.02795	3	=	.11811
9-100	=	.00354	41-100	=	.01614	72-100	=	.02835	4	=	.15748
10-100	=	.00394	42-100	=	.01654	73-100	=	.02874	5	=	.19685
11-100	=	.00433	43-100	=	.01693	74-100	=	.02913	6	=	.23622
12-100	=	.00472	44-100	=	.01732	75-100	=	.02953	7	=	.27559
13-100	=	.00512	45-100	=	.01772	76-100	=	.02992	8	=	.31496
14-100	=	.00551	46-100	=	.01811	77-100	=	.03032	9	=	.35433
15-100	=	.00591	47-100	=	.01850	78-100	=	.03071	10	=	.39370
16-100	=	.00630	48-100	=	.01890	79-100	=	.03110	11	=	.43307
17-100	=	.00669	49-100	=	.01929	80-100	=	.03150	12	=	.47244
18-100	=	.00709	50-100	=	.01969	81-100	=	.03189	13	=	.51181
19-100	=	.00748	51-100	=	.02008	82-100	=	.03228	14	=	.55118
20-100	=	.00787	52-100	=	.02047	83-100	=	.03268	15	=	.59055
21-100	=	.00827	53-100	=	.02087	84-100	=	.03307	16	=	.62992
22-100	=	.00866	54-100	=	.02126	85-100	=	.03346	17	=	.66929
23-100	=	.00906	55-100	=	.02165	86-100	=	.03386	18	=	.70866
24-100	=	.00945	56-100	=	.02205	87-100	=	.03425	19	=	.74803
25-100	=	.00984	57-100	=	.02244	88-100	=	.03465	20	=	.78740
26-100	=	.01024	58-100	=	.02283	89-100	=	.03504	21	=	.82677
27-100	=	.01063	59-100	=	.02323	90-100	=	.03543	22	=	.86614
28-100	=	.01102	60-100	=	.02362	91-100	=	.03583	23	=	.90551
29-100	=	.01142	61-100	=	.02402	92-100	=	.03622	24	=	.94488
30-100	=	.01181	62-100	=	.02441	93-100	=	.03661	25	=	.98425
31-100	=	.01220	63-100	=	.02480	94-100	=	.03701	26	=	1.02362
32-100	=	.01260	....			....			....		

Wire Gauge Standards

Wire gauge no.	Decimal parts of an inch						
	American or Brown & Sharpe	Birmingham or Stubs wire	Washburn & Moen on steel wire gauge	American S. & W. Co.'s music wire	Imperial wire gauge	Stubs steel wire	U.S. standard for plate
0000000	0.651354	....	0.4000	....	0.500	....	0.500
000000	0.580049	....	0.4615	0.004	0.464	....	0.46875

Wire Gauge Standards (Cont'd)

			Decimal parts of an inch				
Wire gauge no.	American or Brown & Sharpe	Birmingham or Stubs wire	Washburn & Moen on steel wire gauge	American S. & W. Co.'s music wire	Imperial wire gauge	Stubs steel wire	U.S. standard for plate
00000	0.516549	0.500	0.4305	0.005	4.432	....	0.43775
0000	0.460	0.454	0.3938	0.006	0.400	....	0.40625
000	0.40964	0.425	0.3625	0.007	0.372	....	0.375
00	0.3648	0.380	0.3310	0.008	0.348	....	0.34375
0	0.32486	0.340	0.3065	0.009	0.324	....	0.3125
1	0.2893	0.300	0.2830	0.010	0.300	0.227	0.28125
2	0.25763	0.284	0.2625	0.011	0.276	0.219	0.265625
3	0.22942	0.259	0.2437	0.012	0.252	0.212	0.250
4	0.20431	0.238	0.2253	0.013	0.232	0.207	0.234375
5	0.18194	0.220	0.2070	0.014	0.212	0.204	0.21875
6	0.16202	0.203	0.1920	0.016	0.192	0.201	0.203125
7	0.14428	0.180	0.1770	0.018	0.176	0.199	0.1875
8	0.12849	0.165	0.1620	0.020	0.160	0.197	0.171875
9	0.11443	0.148	0.1483	0.022	0.144	0.194	0.15625
10	0.10189	0.134	0.1350	0.024	0.128	0.191	0.140625
11	0.090742	0.120	0.1205	0.026	0.116	0.188	0.125
12	0.080808	0.109	0.1055	0.029	0.104	0.185	0.109375
13	0.071961	0.095	0.0915	0.031	0.092	0.182	0.09375
14	0.064084	0.083	0.0800	0.033	0.080	0.180	0.078125
15	0.057068	0.072	0.0720	0.035	0.072	0.178	0.0703125
16	0.05082	0.065	0.0625	0.037	0.064	0.175	0.0625
17	0.045257	0.058	0.0540	0.039	0.056	0.172	0.05625
18	0.040303	0.049	0.0475	0.041	0.048	0.168	0.050
19	0.03589	0.042	0.0410	0.043	0.040	0.164	0.04375
20	0.031961	0.035	0.0348	0.045	0.036	0.161	0.0375
21	0.028462	0.032	0.0317	0.047	0.032	0.157	0.034375
22	0.025347	0.028	0.0286	0.049	0.028	0.155	0.03125
23	0.022571	0.025	0.0258	0.051	0.024	0.153	0.028125
24	0.0201	0.022	0.0230	0.055	0.022	0.151	0.025
25	0.0179	0.020	0.0204	0.059	0.020	0.148	0.021875
26	0.01594	0.018	0.0181	0.063	0.018	0.146	0.01875
27	0.014195	0.016	0.0173	0.067	0.0164	0.143	0.0171875
28	0.012641	0.014	0.0162	0.071	0.0149	0.139	0.015625
29	0.011257	0.013	0.0150	0.075	0.0136	0.134	0.0140625
30	0.010025	0.012	0.0140	0.080	0.0124	0.127	0.0125

Wire Gauge Standards (Cont'd)

			Decimal parts of an inch				
Wire gauge no.	American or Brown & Sharpe	Birmingham or Stubs wire	Washburn & Moen on steel wire gauge	American S. & W. Co.'s music wire	Imperial wire gauge	Stubs steel wire	U.S. standard for plate
31	0.008928	0.010	0.0132	0.085	0.0116	0.120	0.0109375
32	0.00795	0.009	0.0128	0.090	0.0108	0.115	0.01015625
33	0.00708	0.008	0.0118	0.095	0.0100	0.112	0.009375
34	0.006304	0.007	0.0104	....	0.0092	0.110	0.00859375
35	0.005614	0.005	0.0095	....	0.0084	0.108	0.0078125
36	0.005	0.004	0.0090	....	0.0076	0.106	0.00703125
37	0.004453	....	0.0085	....	0.0068	0.103	0.006640625
38	0.003965	....	0.0080	....	0.0060	0.101	0.00625
39	0.003531	....	0.0075	....	0.0052	0.099	
40	0.003144	....	0.0070	....	0.0048	0.097	

Metric Measures

The metric unit of length is the meter = 39.37 inches.

The metric unit of weight is the gram = 15.432 grains.

The following prefixes are used for sub-divisions and multiples: Milli = 1/1000, Centi = 1/100, Deci = 1/10, Deca = 10, Hecto = 100, Kilo = 1000, Myria = 10,000.

Metric and English Equivalent Measures

MEASURES OF LENGTH

Metric		English
1 meter	=	39.37 inches, or 3.28083 feet, or 1.09361 yards
.3048 meter	=	1 foot
1 centimeter	=	.3937 inch
2.54 centimeters	=	1 inch
1 millimeter	=	.03937 inch, or nearly 1-25 inch
25.4 millimeters	=	1 inch
1 kilometer	=	1093.61 yards, or 0.62137 mile

English Conversion Table

Length

Inches	×	.0833	=	feet
Inches	×	.02778	=	yards
Inches	×	.00001578	=	miles
Feet	×	.3333	=	yards
Feet	×	.0001894	=	miles

244

Yards	×	36.00	=	inches
Yards	×	3.00	=	feet
Yards	×	.0005681	=	miles
Miles	×	63360.00	=	inches
Miles	×	5280.00	=	feet
Miles	×	1760.00	=	yards
Circumference of circle	×	.3188	=	diameter
Diameter of circle	×	3.1416	=	circumference

Area

Square inches	×	.00694	=	square feet
Square inches	×	.0007716	=	square yards
Square feet	×	144.00	=	square inches
Square feet	×	.11111	=	square yards
Square yards	×	1296.00	=	square inches
Square yards	×	9.00	=	square feet
Dia. of circle squared	×	.7854	=	area
Dia. of sphere squared	×	3.1416	=	surface

Volume

Cubic inches	×	.0005787	=	cubic feet
Cubic inches	×	.00002143	=	cubic yards
Cubic inches	×	.004329	=	U. S. gallons
Cubic feet	×	1728.00	=	cubic inches
Cubic feet	×	.03704	=	cubic yards
Cubic feet	×	7.4805	=	U. S. gallons
Cubic yards	×	46656.00	=	cubic inches
Cubic yards	×	27.00	=	cubic feet
Dia. of sphere cubed	×	.5236	=	volume

Weight

Grains (avoirdupois)	×	.002286	=	ounces
Ounces (avoirdupois)	×	.0625	=	pounds
Ounces (avoirdupois)	×	.00003125	=	tons
Pounds (avoirdupois)	×	16.00	=	ounces
Pounds (avoirdupois)	×	.01	=	hundredweight
Pounds (avoirdupois)	×	.0005	=	tons
Tons (avoirdupois)	×	32000.00	=	ounces
Tons (avoirdupois)	×	2000.00	=	pounds

Energy

Horsepower	×	33000.	=	ft.-lbs. per min.
B. t. u.	×	778.26	=	ft.-lbs.
Ton of refrigeration	×	200.	=	B. t. u. per min.

Appendix

Pressure

Lbs. per sq. in.	×	2.31	=	ft. of water (60°F.)
Ft. of water (60°F.)	×	.433	=	lbs. per sq. in.
Ins. of water (60°F.)	×	.0361	=	lbs. per sq. in.
Lbs. per sq. in.	×	27.70	=	ins. of water (60°F.)
Lbs. per sq. in.	×	2.041	=	ins. of Hg. (60°F.)
Ins. of Hg (60°F.)	×	.490	=	lbs. per sq. in.

Power

Horsepower	×	746.	=	watts
Watts	×	.001341	=	horsepower
Horsepower	×	42.4	=	B. t. u. per min.

Water Factors (at point of greatest density—39.2°F.)

Miners inch (of water)	×	8.976	=	U. S. gals. per min.
Cubic inches (of water)	×	.57798	=	ounces
Cubic inches (of water)	×	.036124	=	pounds
Cubic inches (of water)	×	.004329	=	U. S. gallons
Cubic inches (of water)	×	.003607	=	English gallons
Cubic feet (of water)	×	62.425	=	pounds
Cubic feet (of water)	×	.03121	=	tons
Cubic feet (of water)	×	7.4805	=	U. S. gallons
Cubic inches (of water)	×	6.232	=	English gallons
Cubic foot of ice	×	57.2	=	pounds
Ounces (of water)	×	1.73	=	cubic inches
Pounds (of water)	×	26.68	=	cubic inches
Pounds (of water)	×	.01602	=	cubic feet
Pounds (of water)	×	.1198	=	U. S. gallons
Pounds (of water)	×	.0998	=	English gallons
Tons (of water)	×	32.04	=	cubic feet
Tons (of water)	×	239.6	=	U. S. gallons
Tons (of water)	×	199.6	=	English gallons
U. S. gallons	×	231.00	=	cubic inches
U. S. gallons	×	.13368	=	cubic feet
U. S. gallons	×	8.345	=	pounds
U. S. gallons	×	.8327	=	English gallons
U. S. gallons	×	3.785	=	liters
English gallons (Imperial)	×	277.41	=	cubic inches
English gallons (Imperial)	×	.1605	=	cubic feet
English gallons (Imperial)	×	10.02	=	pounds
English gallons (Imperial)	×	1.201	=	U. S. gallons
English gallons (Imperial)	×	4.546	=	liters

Metric Conversion Table

Length

Millimeters	×	.03937	= inches
Millimeters	÷	25.4	= inches
Centimeters	×	.3937	= inches
Centimeters	÷	2.54	= Inches
Meters	×	39.37	= inches (Act. Cong.)
Meters	×	3.281	= feet
Meters	×	1.0936	= yards
Kilometers	×	.6214	= miles
Kilometers	÷	1.6093	= miles
Kilometers	×	3280.8	= feet

Area

Sq. Millimeters	×	.00155	= sq. in.
Sq. Millimeters	÷	645.2	= sq. in.
Sq. Centimeters	×	.155	= sq. in.
Sq. Centimeters	÷	6.452	= sq. in.
Sq. Meters	×	10.764	= sq. ft.
Sq. Kilometers	×	247.1	= acres
Hectares	×	2.471	= acres

Volume

Cu. Centimeters	÷	16.387	= cu. in.
Cu. Centimeters	÷	3.69	= fl. drs. (U.S.P.)
Cu. Centimeters	÷	29.57	= fl. oz. (U.S.P.)
Cu. Meters	×	35.314	= cu. ft.
Cu. Meters	×	1.308	= cu. yards
Cu. Meters	×	264.2	= gals. (231 cu. in.)
Litres	×	61.023	= cu. in. (Act. Cong.)
Litres	×	33.82	= fl. oz. (U.S.J.)
Litres	×	.2642	= gals. (231 cu. in.)
Litres	÷	3.785	= gals. (231 cu. in.)
Litres	÷	28.317	= cu. ft.
Hectolitres	×	3.531	= cu. ft.
Hectolitres	×	2.838	= bu. (2150.42 cu. in.)
Hectolitres	×	.1308	= cu. yds.
Hectolitres	×	26.42	= gals. (231 cu. in.)

Weight

Grams	×	15.432	= grains (Act. Cong.)
Grams	÷	981.	= dynes
Grams (water)	÷	29.57	= fl. oz.
Grams	÷	28.35	= oz. avoirdupois
Kilo-grams	×	2.2046	= lbs.

The Audel® Mail Order Bookstore

Here's an opportunity to order the valuable books you may have missed before and to build your own personal, comprehensive library of Audel books. You can choose from an extensive selection of technical guides and reference books. They will provide access to the same sources the experts use, put all the answers at your fingertips, and give you the know-how to complete even the most complicated building or repairing job, in the same professional way.

Each volume:
- ● **Fully illustrated**
- ● **Packed with up-to-date facts and figures**
- ● **Completely indexed for easy reference**

APPLIANCES

HOME APPLIANCE SERVICING, 3rd Edition
A practical book for electric & gas servicemen, mechanics & dealers. Covers the principles, servicing, and repairing of home appliances. 592 pages; 5¼ x 8¼; hardbound. Price: $12.95

REFRIGERATION AND AIR CONDITIONING LIBRARY—2 Vols. Price: $21.95

REFRIGERATION: HOME AND COMMERCIAL
Covers the whole realm of refrigeration equipment from fractional-horsepower water coolers, through domestic refrigerators to multi-ton commercial installations. 656 pages; 5½ x 8¼; hardbound. Price: $12.95

AIR CONDITIONING: HOME AND COMMERCIAL
A concise collection of basic information, tables, and charts for those interested in understanding, troubleshooting, and repairing home air conditioners and commercial installations. 464 pages; 5½ x 8¼; hardbound. Price: $10.95

OIL BURNERS, 3rd Edition
Provides complete information on all types of oil burners and associated equipment. Discusses burners—blowers—ignition transformers—electrodes—nozzles—fuel pumps—filters—Controls. Installation and maintenance are stressed. 320 pages; 5½ x 8¼; hardbound. Price: $9.95

Use the order coupon on the back page of this book.

All prices are subject to change without notice.

AUTOMOTIVE

AUTOMOBILE REPAIR GUIDE, 4th Edition
A practical reference for auto mechanics, servicemen, trainees, and owners Explains theory, construction, and servicing of modern domestic motorcars. 800 pages; 5½ x 8¼; hardbound. **Price: 14.95**

CAN-DO TUNE-UP™ SERIES
Each book in this series comes with an audio tape cassette. Together they provide an organized set of instructions that will show you and talk you through the maintenance and tune-up procedures designed for your particular car. All books are softcover.

AMERICAN MOTORS CORPORATION CARS
(The 1964 thru 1974 cars covered include: Matador. Rambler. Gremlin. and AMC Jeep (Willys).) 112 pages; 5½ x 8½; softcover. **Price: $7.95**

CHRYSLER CORPORATION CARS
(The 1964 thru 1974 cars covered include: Chrysler, Dodge, and Plymouth.) 112 pages; 5½ x 8½; softcover. **Price: $7.95**

FORD MOTOR COMPANY CARS
(The 1954 thru 1974 cars covered include: Ford, Lincoln, and Mercury.) 112 pages; 5½ x 8½; softcover. **Price: $7.95**

GENERAL MOTORS CORPORATION CARS
(The 1964 thru 1974 cars covered include: Buick, Cadillac, Chevrolet, Oldsmobile, and Pontiac.) 112 pages; 5½ x 8½; softcover. **Price: $7.95**

PINTO AND VEGA CARS
1971 thru 1974. 112 pages. 5½ x 8½; softcover. **Price: $7.95**

TOYOTA AND DATSUN CARS
1964 thru 1974. 112 pages; 5½ x 8½; softcover. **Price: $7.95**

VOLKSWAGEN CARS
(The 1964 thru 1974 cars covered include: Beetle. Super Beetle. and Karmann Ghia.) 96 pages; 5½ x 8½; softcover. **Price: $7.95**

AUTOMOTIVE AIR CONDITIONING
You can easily perform most all service procedures you've been paying for in the past. This book covers the systems built by the major manufacturers, even after-market installations. Contents: introduction—refrigerant—tools—air conditioning circuit—general service procedures—electrical systems—the cooling system—system diagnosis—electrical diagnosis—troubleshooting. 232 pages; 5½ x 8½; softcover. **Price: $7.95**

Use the order coupon on the back page of this book.

All prices are subject to change without notice.

DIESEL ENGINE MANUAL, 3rd Edition

A practical guide covering the theory, operation, and maintenance of modern diesel engines. Explains diesel principles—valves—timing—fuel pumps—pistons and rings—cylinders—lubrication —cooling system—fuel oil and more. 480 pages; 5½ x 8¼; hardbound. **Price: $10.95**

GAS ENGINE MANUAL, 2nd Edition

A completely practical book covering the construction, operation, and repair of all types of modern gas engines. 400 pages; 5½ x 8¼; hardbound. **Price: $9.95**

BUILDING AND MAINTENANCE

ANSWERS ON BLUEPRINT READING, 3rd Edition

Covers all types of blueprint reading for mechanics and builders. This book reveals the secret language of blueprints, step-by-step in easy stages. 312 pages; 5½ x 8¼; hardbound. **Price: $9.95**

BUILDING MAINTENANCE, 2nd Edition

Covers all the practical aspects of building maintenance. Painting and decorating; plumbing and pipe fitting; carpentry; heating maintenance; custodial practices and more. (A book for building owners, managers, and maintenance personnel.) 384 pages; 5½ x 8¼; hardbound. **Price: $9.95**

COMPLETE BUILDING CONSTRUCTION

At last—a *one-volume* instruction manual to show you how to construct a frame or brick building from the footings to the ridge. Build your own garage, tool shed, other outbuilding—even your own house or place of business. Building construction tells you how to lay out the building and excavation lines on the lot; how to make concrete forms and pour the footings and foundation; how to make concrete slabs, walks, and driveways; how to lay concrete block, brick and tile; how to build your own fireplace and chimney: It's one of the newest Audel books, clearly written by experts in each field and ready to help you every step of the way. 800 pages; 5½ x 8¼; hardbound. **Price: 19.95**

GARDENING & LANDSCAPING

A comprehensive guide for homeowners and for industrial, municipal, and estate groundskeepers. Gives information on proper care of annual and perennial flowers; various house plants; greenhouse design and construction; insect and rodent controls; and more. 384 pages; 5½ x 8¼; hardbound. **Price: $9.95**

CARPENTERS & BUILDERS LIBRARY, 4th Edition (4 Vols.)

A practical, illustrated trade assistant on modern construction for carpenters, builders, and all woodworkers. Explains in practical, concise language and illustrations all the principles, advances, and shortcuts based on modern practice. How to calculate various jobs. **Price: $35.95**

> Vol. 1—Tools, steel square, saw filing, joinery cabinets. 384 pages; 5½ x 8¼; hardbound. **Price: $10.95**
> Vol. 2—Mathematics, plans, specifications, estimates 304 pages; 5½ x 8¼; hardbound. **Price: $10.95**
> Vol. 3—House and roof framing, laying out foundations. 304 pages; 5½ x 8¼; hardbound. **Price: $10.95**
> Vol. 4—Doors, windows, stairs, millwork, painting. 368 pages; 5½ x 8¼; hardbound. **Price: $10.95**

Use the order coupon on the back page of this book.

All prices are subject to change without notice.

CARPENTRY AND BUILDING

Answers to the problems encountered In today's building trades. The actual questions asked of an architect by carpenters and builders are answered in this book. 448 pages; 5½ x 8¼; hardbound. **Price: $10.95**

WOOD STOVE HANDBOOK

The wood stove handbook shows how wood burned in a modern wood stove offers an immediate, practical, low-cost method of full-time or part-time home heating. The book points out that wood is plentiful, low in cost (sometimes free), and nonpolluting, especially when burned in one of the newer and more efficient stoves. In this book, you will learn about the nature of heat and its control, what happens inside and outside a stove, how to have a safe and efficient chimney, and how to install a modern wood burning stove. You will also learn about the different types of firewood and how to get it, cut it, split it, and store it. 128 pages; 8½ x 11; softcover. **Price: $7.95**

HEATING, VENTILATING, AND AIR CONDITIONING LIBRARY (3 Vols.)

This three-volume set covers all types of furnaces, ductwork, air conditioners, heat pumps, radiant heaters, and water heaters, including swimming-pool heating systems. **Price: $32.95**

Volume 1

Partial Contents: Heating Fundamentals . . . Insulation Principles . . . Heating Fuels . . . Electric Heating System . . . Furnace Fundamentals . . . Gas-Fired Furnaces . . . Oil-Fired Furnaces . . . Coal-Fired Furnaces . . . Electric Furnaces. **Price: $11.95**

Volume 2

Partial Contents: Oil Burners . . . Gas Burners . . . Thermostats and Humidistats . . . Gas and Oil Controls . . . Pipes, Pipe Fitting, and Piping Details . . . Valves and Valve Installations 560 pages; 5½ x 8¼; hardbound. **Price: $11.95**

Volume 3

Partial Contents: Radiant Heating . . . Radiators, Convectors, and Unit Heaters . . . Stoves, Fireplaces, and Chimneys . . . Water Heaters and Other Appliances . . . Central Air Conditioning Systems . . . Humidifiers and Dehumidifiers. 544 pages; 5½ x 8¼; hardbound. **Price: $11.95**

HOME MAINTENANCE AND REPAIR: Walls, Ceilings, and Floors

Easy-to-follow instructions for sprucing up and repairing the walls, ceiling, and floors of your home. Covers nail pops, plaster repair, painting, paneling, ceiling and bathroom tile, and sound control, 80 pages; 8½ x 11; softcover. **Price: $6.95**

HOME PLUMBING HANDBOOK , 2nd Edition

A complete guide to home plumbing repair and installation. 200 pages; 8½ x 11; softcover. **Price: $7.95**

MASONS AND BUILDERS LIBRARY—2 Vols.

A practical, illustrated trade assistant on modern construction for bricklayers, stone-masons, cement workers, plasterers, and tile setters. Explains all the principles, advances, and shortcuts based on modern practice—including how to figure and calculate various jobs. **Price: $17.95**

Vol. 1—Concrete, Block, Tile, Terrazzo. 368 pages; 5½ x 8¼; hardbound. **Price: $9.95**

Vol. 2—Bricklaying, Plastering, Rock Masonry, Clay Tile. 384 pages; 5½ x 8¼; hardbound. **Price: 9.95**

Use the order coupon on the back page of this book.

All prices are subject to change without notice.

PLUMBERS AND PIPE FITTERS LIBRARY—3 Vols.

A practical, illustrated trade assistant and reference for master plumbers, journeymen and apprentice pipe fitters, gas fitters and helpers, builders, contractors, and engineers. Explains in simple language, illustrations, diagrams, charts, graphs, and pictures, the principles of modern plumbing and pipe-fitting practices. **Price: $26.95**

Vol. 1—Materials, tools, roughing-in. 320 pages; 5½ x 8¼; hardbound. **Price: $9.95**

Vol. 2—Welding, heating, air-conditioning. 384 pages; 5½ x 8¼; hardbound. **Price: $9.95**

Vol. 3—Water supply, drainage, calculations. 272 pages; 5½ x 8¼; hardbound. **Price: $9.95**

PLUMBERS HANDBOOK

A pocket manual providing reference material for plumbers and/or pipe fitters. General information sections contain data on cast-iron fittings, copper drainage fittings, plastic pipe, and repair of fixtures. 288 pages; 4 x 6; softcover. **Price: $9.95**

QUESTIONS AND ANSWERS FOR PLUMBERS EXAMINATIONS, 2nd Edition

Answers plumbers' questions about types of fixtures to use, size of pipe to install, design of systems, size and location of septic tank systems, and procedures used in installing material. 256 pages; 5½ x 8¼; softcover. **Price: $8.95**

TREE CARE MANUAL

The conscientious gardener's guide to healthy, beautiful trees. Covers planting, grafting, fertilizing, pruning, and spraying. Tells how to cope with insects, plant diseases, and environmental damage. 224 pages; 8½ x 11; softcover. **Price: $8.95**

UPHOLSTERING

Upholstering is explained for the average householder and apprentice upholsterer. From repairing and reglueing of the bare frame, to the final sewing or tacking, for antiques and most modern pieces, this book covers it all. 400 pages; 5½ x 8¼; hardbound. **Price: $9.95**

WOOD FURNITURE: Finishing, Refinishing, Repairing

Presents the fundamentals of furniture repair for both veneer and solid wood. Gives complete instructions on refinishing procedures, which includes stripping the old finish, sanding, selecting the finish and using wood fillers. 352 pages; 5½ x 8¼; hardbound. **Price: $9.95**

ELECTRICITY/ELECTRONICS

ELECTRICAL LIBRARY

If you are a student of electricity or a practicing electrician, here is a very important and helpful library you should consider owning. You can learn the basics of electricity, study electric motors and wiring diagrams, learn how to interpret the NEC, and prepare for the electrician's examination by using these books.

Electric Motors, 3rd Edition. 528 pages; 5½ x 8¼; hardbound. **Price: $10.95**

Guide to the 1981 National Electrical Code. 608 pages; 5½ x 8¼; hardbound. **Price: $13.95**

House Wiring, 5th Edition. 256 pages; 5½ x 8¼; hardbound. **Price: $9.95**

Practical Electricity, 3rd Edition. 496 pages; 5½ x 8¼; hardbound. **Price: $10.95**

Questions and Answers for Electricians Examinations, 7th Edition. 288 pages; 5½ x 8¼; hardbound. **Price: $9.95**

ELECTRICAL COURSE FOR APPRENTICES AND JOURNEYMEN

A study course for apprentice or journeymen electricians. Covers electrical theory and its applications. 448 pages; 5½ x 8¼; hardbound. **Price: $10.95**

Use the order coupon on the back page of this book.

All prices are subject to change without notice.

RADIOMANS GUIDE, 4th Edition

Contains the latest information on radio and electronics from the basics through transistors. 480 pages; 5½ x 8¼; hardbound. **Price: $11.95**

TELEVISION SERVICE MANUAL, 4th Edition

Provides the practical information necessary for accurate diagnosis and repair of both black-and-white and color television receivers. 512 pages; 5½ x 8¼; hardbound. **Price: $11.95**

ENGINEERS/MECHANICS/ MACHINISTS

MACHINISTS LIBRARY, 3rd Edition

Covers modern machine-shop practice. Tells how to set up and operate lathes, screw and milling machines, shapers, drill presses, and all other machine tools. A complete reference library. **Price: $29.95**

Vol. 1—Basic Machine Shop. 352 pages; 5½ x 8¼; hardbound. **Price: $10.95**

Vol. 2—Machine Shop. 480 pages; 5½ x 8¼; hardbound. **Price: $10.95**

Vol. 3—Toolmakers Handy Book. 400 pages; 5½ x 8¼; hardbound. **Price: $10.95**

MECHANICAL TRADES POCKET MANUAL

Provides practical reference material for mechanical tradesmen. This handbook covers methods, tools, equipment, procedures, and much more. 256 pages; 4 x 6; softcover. **Price: $8.95**

MILLWRIGHTS AND MECHANICS GUIDE, 2nd Edition

Practical information on plant installation, operation, and maintenance for millwrights, mechanics, maintenance men, erectors, riggers, foremen, inspectors, and superintendents. 960 pages; 5½ x 8¼; hardbound. **Price: $16.95**

POWER PLANT ENGINEERS GUIDE, 2nd Edition

The complete steam or diesel power-plant engineer's library. 816 pages; 5½ x 8¼; hardbound. **Price: $15.95**

QUESTIONS AND ANSWERS FOR ENGINEERS AND FIREMANS EXAMINATIONS, 3RD EDITION

Presents both legitimate and "catch" questions with answers that may appear on examinations for engineers and firemans licenses for stationary, marine, and combustion engines. 496 pages; 5½ x 8¼; hardbound. **Price: $10.95**

WELDERS GUIDE, 2nd Edition

This new edition is a practical and concise manual on the theory, practical operation, and maintenance of all welding machines. Fully covers both electric and oxy-gas welding. 928 pages; 5½ x 8¼; hardbound. **Price: $14.95**

WELDER/FITTERS GUIDE

Provides basic training and instruction for those wishing to become welder/fitters. Step-by-step learning sequences are presented from learning about basic tools and aids used in weldment assembly, through simple work practices, to actual fabrication of weldments. 160 pages· 8½ x 11; softcover. **Price: $7.95**

Use the order coupon on the back page of this book.

All prices are subject to change without notice.

FLUID POWER

PNEUMATICS AND HYDRAULICS, 3rd Edition

Fully discusses installation, operation, and maintenance of both HYDRAULIC AND PNEUMATIC (air) devices. 496 pages; 5½ x 8¼; hardbound. **Price: $10.95**

PUMPS, 3rd Edition

A detailed book on all types of pumps from the old-fashioned kitchen variety to the most modern types. Covers construction, application, installation, and troubleshooting. 480 pages; 5½ x 8¼; hardbound. **Price: $10.95**

HYDRAULICS FOR OFF-THE-ROAD EQUIPMENT

Everything you need to know from basic hydraulics to troubleshooting hydraulic systems on off-the-road equipment. Heavy-equipment operators, farmers, fork-lift owners and operators, mechanics—all need this practical, fully illustrated manual. 272 pages; 5½ x 8¼; hardbound. **Price: $8.95**

HOBBY

COMPLETE COURSE IN STAINED GLASS

Written by an outstanding artist in the field of stained glass, this book is dedicated to all who love the beauty of the art. Ten complete lessons describe the required materials, how to obtain them, and explicit directions for making several stained glass projects. 80 pages; 8½ x 11; softbound. **Price: $6.95**

BUILD YOUR OWN AUDEL
DO-IT-YOURSELF LIBRARY AT HOME!

Use the handy order coupon today to gain the valuable information you need in all the areas that once required a repairman. Save money and have fun while you learn to service your own air conditioner, automobile, and plumbing. Do your own professional carpentry, masonry, and wood furniture refinishing and repair. Build your own security systems. Find out how to repair your TV or Hi-Fi. Learn landscaping, upholstery, electronics and much, much more.

All prices are subject to change without notice.

HERE'S HOW TO ORDER

1. Enter the correct title(s) and author(s) of the book(s) you want in the space(s) provided.

2. Print your name, address, city, state and zip code clearly.

3. Detach the order coupon below and mail today to:

Theodore Audel & Company
4300 West 62nd Street
Indianapolis, Indiana 46206
ATTENTION: ORDER DEPT.

All prices are subject to change without notice.

- -

ORDER COUPON

Please rush the following book(s).

Title_____

Author_____

Title_____

Author_____

NAME_____

ADDRESS_____

CITY_____ STATE_____ ZIP_____

☐ Payment enclosed _____
   (No shipping and   Total
   handling charge)

☐ Bill me (shipping and handling charge will be added)
Add local sales tax where applicable.

Litho in U.S.A.

HERE'S HOW TO ORDER

Select the Audel book(s) you want, fill in the order card below, detach and mail today. Send no money now. You'll have 15 days to examine the books in the comfort of your own home. If not completely satisfied, simply return your order and owe nothing.

If you decide to keep the books, we will bill you for the total amount, plus a small charge for shipping and handling.

1. Enter the correct title(s) and author(s) of the book(s) you want in the space(s) provided.

2. Print your name, address, city, state and zip code clearly.

3. Detach the order card below and mail today. No postage is required.

Detach postage-free order card on perforated line

FREE TRIAL ORDER CARD

☐ Please rush the following book(s) for my free trial. I understand if I'm not completely satisfied, I may return my order within 15 days and owe nothing. Otherwise, you will bill me for the total amount plus a small postage & handling charge.

Title_____

Author_____

Title_____

Author_____

NAME_____

ADDRESS_____

CITY_____ STATE_____ ZIP_____

☐ Save postage & handling costs. Full payment enclosed (plus sales tax, if any).

Cash must accompany orders under $5.00.
Money-back guarantee still applies.

**DETACH POSTAGE-PAID REPLY CARD
BELOW AND MAIL TODAY!**

Just select your books, enter the titles and
authors on the order card, fill out your name
and address, and mail. There's no need to
send money.

15-Day Free Trial On All Books . . .